Movable Bed Physical Models

NATO ASI Series

Advanced Science Institutes Series

A Series presenting the results of activities sponsored by the NATO Science Committee, which aims at the dissemination of advanced scientific and technological knowledge, with a view to strengthening links between scientific communities.

The Series is published by an international board of publishers in conjunction with the NATO Scientific Affairs Division

A Life Sciences **B Physics**	Plenum Publishing Corporation London and New York
C Mathematical and Physical Sciences **D Behavioural and Social Sciences** **E Applied Sciences**	Kluwer Academic Publishers Dordrecht, Boston and London
F Computer and Systems Sciences **G Ecological Sciences** **H Cell Biology**	Springer-Verlag Berlin, Heidelberg, New York, London, Paris and Tokyo

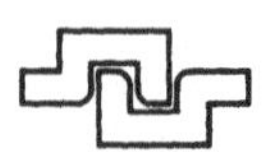

Series C: Mathematical and Physical Sciences - Vol. 312

Movable Bed Physical Models

edited by

Hsieh Wen Shen

Department of Civil Engineering,
University of California,
Berkeley, CA, U.S.A.

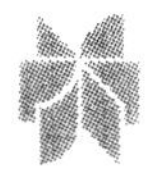

Kluwer Academic Publishers

Dordrecht / Boston / London

Published in cooperation with NATO Scientific Affairs Division

Proceedings of the NATO Advanced Research Workshop on
Movable Bed Physical Models
Delft, The Netherlands
August 18–21, 1987

Library of Congress Cataloging in Publication Data

ISBN 0-7923-0828-X

Published by Kluwer Academic Publishers,
P.O. Box 17, 3300 AA Dordrecht, The Netherlands.

Kluwer Academic Publishers incorporates the publishing programmes of
D. Reidel, Martinus Nijhoff, Dr W. Junk and MTP Press.

Sold and distributed in the U.S.A. and Canada
by Kluwer Academic Publishers,
101 Philip Drive, Norwell, MA 02061, U.S.A.

In all other countries, sold and distributed
by Kluwer Academic Publishers Group,
P.O. Box 322, 3300 AH Dordrecht, The Netherlands.

Printed on acid-free paper

Printed in the Netherlands

This book contains the proceedings of a NATO Advanced Research Workshop held within the programme of activities of the NATO Special Programme on Global Transport Mechanisms in the Geo-Sciences running from 1983 to 1988 as part of the activities of the NATO Science Committee.

Other books previously published as a result of the activities of the Special Programme are:

BUAT-MENARD, P. (Ed.) – *The Role of Air-Sea Exchange in Geochemical Cycling* (C185) 1986

CAZENAVE, A. (Ed.) – *Earth Rotation: Solved and Unsolved Problems* (C187) 1986

WILLEBRAND, J. and ANDERSON, D.L.T. (Eds.) – *Large-Scale Transport Processes in Oceans and Atmosphere* (C190) 1986

NICOLIS, C. and NICOLIS, G. (Eds.) – *Irreversible Phenomena and Dynamical Systems Analysis in Geosciences* (C192) 1986

PARSONS, I. (Ed.) – *Origins of Igneous Layering* (C196) 1987

LOPER, E. (Ed.) – *Structure and Dynamics of Partially Solidified Systems* (E125) 1987

VAUGHAN, R. A. (Ed.) – *Remote Sensing Applications in Meteorology and Climatology* (C201) 1987

BERGER, W. H. and LABEYRIE, L. D. (Eds.) – *Abrupt Climatic Change – Evidence and Implications* (C216) 1987

VISCONTI, G. and GARCIA, R. (Eds.) – *Transport Processes in the Middle Atmosphere* (C213) 1987

SIMMERS, I. (Ed.) – *Estimation of Natural Recharge of Groundwater* (C222) 1987

HELGESON, H. C. (Ed.) – *Chemical Transport in Metasomatic Processes* (C218) 1987

CUSTODIO, E., GURGUI, A. and LOBO FERREIRA, J. P. (Eds.) – *Groundwater Flow and Quality Modelling* (C224) 1987

ISAKSEN, I. S. A. (Ed.) – *Tropospheric Ozone* (C227) 1988

SCHLESINGER, M.E. (Ed.) – *Physically-Based Modelling and Simulation of Climate and Climatic Change 2 vols.* (C243) 1988

UNSWORTH, M. H. and FOWLER, D. (Eds.) – *Acid Deposition at High Elevation Sites* (C252) 1988

KISSEL, C. and LAY, C. (Eds.) – *Paleomagnetic Rotations and Continental Deformation* (C254) 1988

HART, S. R. and GULEN, L. (Eds.) – *Crust/Mantle Recycling at Subduction Zones* (C258) 1989

GREGERSEN, S. and BASHAM, P. (Eds.) – *Earthquakes at North-Atlantic Passive Margins: Neotectonics and Postglacial Rebound* (C266) 1989

MOREL-SEYTOUX, H. J. (Ed.) – *Unsaturated Flow in Hydrologic Modeling* (C275) 1989

BRIDGWATER, D. (Ed.) – *Fluid Movements – Element Transport and the Composition of the Crust* (C281) 1989

LEINEN, M. and SARNTHEIN, M. (Eds.) – *Paleoclimatology and Paleometeorology: Modern and Past Patterns of Global Atmospheric Transport* (C282) 1989

ANDERSON, D.L.T. and WILLEBRAND, J. (Eds.) – *Ocean Circulation Models: Combining Data and Dynamics* (C284) 1989

BERGER, A., SCHNEIDER, S. and DUPLESSY, J. Cl. (Eds.) – *Climate and Geo-Sciences* (C285) 1989

KNAP, A.H. (Ed.) – *The Long-Range Atmospheric Transport of Natural and Contaminant Substances from Continent to Ocean and Continent to Continent* (C297) 1990.

BLEIL, U. and THIEDE, J. (Eds.) – *Geological History of the Polar Oceans – Arctic Versus Antarctic* (C308) 1990.

Table of Contents

PREFACE

For centuries, physical models have been used to investigate complex hydraulic problems. Leonardo da Vinci (1452-1519) stated, "I will treat of such a subject. But first of all, I shall make a few experiments and then demonstrate why bodies are forced to act in this matter." Even with the current advancements of mathematical numerical models, certain complex three-dimensional flow phenomena must still rely on physical model studies. Mathematical models cannot provide adequate solutions if physical processes involved are not completely known. Physical models are particularly attractive to investigate phenomena-involved sediment movements because many three-dimensional sediment processes are still unclear at this stage.

Theoretically, there are numerous factors governing movable bed processes and it is nearly impossible to design model studies to obey all the model criteria. Sometimes, appropriate lightweight materials are difficult or too costly to obtain. Often, distorted models are used due to the limitation of available space and the requirement for greater vertical flow depth to investigate vertical differences of various parameters. The turbulence level in the model may also be maintained at a sufficient level to reproduce a similar flow pattern in the prototype. Frequently, engineers are forced to employ distorted models that cannot be designed to satisfy all governing criteria correctly. Thus each hydraulic laboratory has developed its own rules for model testing and a great deal of experience is needed to interpret model results. This NATO conference on movable bed models was designed to exchange experiences from specialists who have conducted various phases of movable bed studies. Thus all participants were required to submit papers describing their experiences on this subject. Active discussions were held at the later part of the conference to exchange knowledge. All participants felt they benefited from this participation.

This book contains the papers presented in the NATO Workshop on Movable Bed Models. The papers have been arranged according to the following principles. The first two papers give an overview of the state of knowledge and the difficulties encountered in using movable bed models. The next ten papers discuss the experiences and/or analysis of movable bed models for certain types of problems. The last three papers present some special knowledge and problems.

This workshop was organized by a special committee which includes Professor Miguel Countinbo of the University of Portugual; Mr. G. J. Klaassen of the Delft De Voorst Hydraulic Laboratory, The Netherlands; Professor H. Kobus of the University of Stuttgart, West Germany; Professor M.S. Yalin of the Queens University, Canada and myself. We particulary wish to express our sincere gratitude to Mr. Klaassen and the Delft De Voorst Laboratory for their efficient arrangements for the conference. We are indepted to NATO for their sponsorship through the NATO Science Committee Program on Global Transport in Geo-Sciences. Dr. Luis da Cunha has provided much useful advice. Ms. Joanne Birdsall from the University of California at Berkeley assisted in preparing the manuscripts for publication.

Hsieh Wen Shen
University of California
Director, NATO Workshop on Movable Bed Models
January 1990

INTRODUCTORY REMARKS FOR THE NATO WORKSHOP ON MOVABLE BED PHYSICAL MODELS

HSIEH WEN SHEN
University of California
Hydraulic and Coastal Engineering Department
412 O'Brien Hall
Berkeley, CA 94720
U.S.A.

Movable bed models have long been employed to investigate complex hydraulic phenomena. However, the design and usage of models still greatly rely upon experiments and judgments. The interpretation of results from the famous movable bed tests for the Yellow River in China, conducted by the eminent Professor Herbert Engels, is yet to be agreed upon. We wish to thank NATO for partially sponsoring this Workshop so that researchers from different countries can share knowledge on this important subject.

The main functions of a physical model may include: 1) the duplication of selected known flow phenomena in a river; 2) the examination of the effects of certain selected factors on specific flow phenomena; and 3) the investigations and predictions of the resultant changes of flow phenomena for different ranges of flow parameters. Since this paper serves as an introduction to this conference, some existing knowledge on movable bed models will be briefly reviewed. This presentation will suggest questions for discussion in this workshop.

1. SIMILARITY CRITERIA FROM THE BASIC EQUATION OF MOTION

For an incompressible fluid in a rigid boundary, the Navier–Stokes equations are:

H. W. Shen (ed.), Movable Bed Physical Models, 1–12.

$$\frac{Du_i}{Dt} = -g\frac{\partial h}{\partial x_i} - \frac{1\partial p}{\rho\partial x_i} + \frac{\mu}{\rho}\nabla^2 u_i \tag{1}$$

where u_i is the flow velocities in x_i direction, t is the time, h is the flow depth, p is the pressure, g is the gravitational constant, μ is the viscosity and ρ is the fluid density.

The above equation can be written with a dimensionless form as:

$$\frac{Du_i^0}{Dt} = \frac{1}{F}\frac{\partial h^0}{\partial x_i^0} - \frac{\partial p^0}{\partial x_i^0} + \frac{1}{R}\nabla^2 u_i^0 \tag{2}$$

where $u_i^0 = u_i/V_0$, F = Froude number, R = Reynolds number, $h^0 = h/L$, V_0 is a characteristic flow velocity, and L is a characteristic flow length. Equation 2 indicates that the two flow systems are similar if and only if both the Froude number and the Reynolds number are the same in both flow systems. Thus, both of those numbers must be the same for model and prototype if the Navier-Stokes equation is the governing equation involved.

Let the subscript r denote the ratio between that quantity for the prototype and the same quantity for the model; then the conditions for the same Reynolds number and Froude number require the following (where υ is the kinematic viscosity, V is the flow velocity, and L is the length):

$$\frac{V_r L_r}{\nu_r} = 1 = \frac{V_r}{\sqrt{g_r L_r}} \tag{3}$$

$$\nu_r = \frac{L_r^{3/2}}{g_r^{1/2}} \tag{4}$$

It is difficult to select a fluid in the model with the proper ν_r as required by Equation 4 and thus, in most physical model studies, either the Reynolds criterion or the Froude criterion is ignored. For rigid bed river models, gravitational forces are assumed to be important and the Reynolds number is ignored in the main equation. Since resistance is important for the verification of the physical model, a certain range of Reynolds numbers may be required in the model. Since this paper is mainly oriented toward movable bed models, no further discussion is presented for rigid bed models. For movable bed processes, there is not a single governing equation (or a set of equations) that can be found to determine the variation of alluvial bed resistance, suspended sediment load, bed load, channel width, etc., and thus the modeling procedure involves the selection of different governing parameters (not governing equations) for different problems. Usually it is difficult to design a model according to all governing parameters. Similar to the problem discussed in the previous section, certain parameters may be ignored and/or adjusted.

2. SELECTION OF GOVERNING PARAMETERS

2.1 Movements of Bed

In order to study sediment movements in a movable bed river model, the most important criterion is the movement of sediment particles on the beds. Numerous bed load transport rate equations have been proposed. According to Chien (1980), most of the bed load equations can be converted into the following Einstein's bed intensity parameter (the inverse of the Shields number):

$$\pi_1 = \frac{\tau_r}{(\gamma_s - \gamma)_r d_r} \tag{5}$$

where τ is the bed shear stress; γ_s and γ are respectively the specific weights of sediment and fluid; and d is the sediment particle size.

If the main interest is not to study bed load movement rate and the only concern is to create a movable bed, one may neglect the use of π_1 as given in Equation 5 and only to have $\pi_1 > 0.05$ in the model for R_* greater than 100, where $R_* = u_* d_{50}/\nu$. u_* is defined as the shear velocity and is equal to $\sqrt{\tau/\rho}$ where ρ is the fluid density.

The time scale for bed load motion can be calculated as follows:

$$\frac{\partial q_{sb}}{\partial x} = -\gamma_0 \frac{\partial y}{\partial t} \tag{6}$$

where q_{sb} is the bed load transport rate, γ_0 is the porosity, y is the bed level, and t is the time. From Equation 6 the time scale for bed load can be expressed as:

$$t_{br} = \frac{\gamma_{or} Y_r X_r}{q_{sbr}} \tag{7}$$

2.2 Friction Criterion and Bed Forms

The second important factor is the friction factor. Actually, this factor is almost as important as the first factor on movement of bed. Most of the efforts for model calibration is to duplicate the friction factor in the prototype so that the model can reproduce similar friction effects as those in the river.

Usually the two friction equations used are Chezy's equation and Manning's equation, as follows:

$$V = C\sqrt{g}\sqrt{RS} \tag{8}$$

where C is Chezy's friction factor, R is the hydraulic radius, and S is the energy slope.

$$V = \frac{1}{n} R^{2/3} S^{1/2} \tag{9}$$

Einstein and Chien (1956) proposed a generalized friction equation:

$$V = \frac{C_0 \sqrt{g}}{d^m} S^{1/2} R^{(1/2+m)} \tag{10}$$

where m can be adjusted. Equation 10 is identical to Equation 9, if $m = 1/6$ and $n \sim d^{1/6}$. The relationship between C_0 and n is expressed by the following relationship:

$$C_0 = \frac{1}{\sqrt{g}} \frac{d^{1/6}}{n} \tag{11}$$

The Nanjing Institute of Hydraulic Research in China (1978) proposed using relative roughness to express the above friction factor in the following manner:

$$\frac{V}{u_*} = \frac{1}{\kappa} \ln\left(11 \frac{R}{k_s}\right) \tag{12}$$

where k_s is the absolute bed roughness height and κ is the Von Karman's "constant." Combining the definition of u_*, Equations 11 and 12, one obtains (with $\kappa = 0.4$):

$$C_0 = 2.5 \ln\left(11 \frac{R}{k_s}\right) \tag{13}$$

The Nanjing Research Institute suggested using $k_s = d_{50}$ for sediment bed material size greater than d_{50} and $k_s = 0.5$ mm for sediment bed material size less than 0.5 mm. They also recommended using the Manning equation to estimate roughness value. Wang (1984) stated that results from experiments conducted in Tientsin, China, with $1.45 < D/d < 17.1$; 5 cm $< h <$ 10 cm; and $0.15 < V < 0.3$ m/sec; where D is the distance (both longitudinally and transversely) between two artificial roughnesses and h is the flow depth, indicated that κ_n can be expressed as a function of D/d, where:

$$n = \kappa_n d^{1/6} \tag{14}$$

This type of information is rather useful (mainly for a rigid bed) because one can determine the approximate size of roughness needed in the physical model for model calibration.

Often, increasing the roughness of the model surface is required. The commonly used methods are: roughen the surface of the bed and the bank by (1) inserting gravel into the concrete surface; or (2) fastening copper wires and tubes to the concrete surface. Occasionally the model is too small and either the slope and/or the flow velocity must be further adjusted according to the following tow criteria.

According to Manning's equation:

$$V_r = \frac{1}{n_r} R_r^{2/3} S_r^{1/2} \tag{9}$$

The slope can be adjusted as follows:

$$S_r^{1/2} = \frac{V_R n_r}{R_r^{2/3}} = \frac{X_r^{1/2} n_r}{X_r^{2/3}} = \frac{n_r}{X_r^{1/6}} \tag{15}$$

where $V_r = Xr^{1/2}$ according to Froude's criterion and it is for an undistorted model. The flow velocity can be adjusted as follows: $(S_r = 1)$

$$V_r = \frac{X_r^{2/3}}{n_r} \tag{16}$$

For a movable bed model, it is difficult to simulate the field friction factor in the model. A common practice is to use gravel (either movable or rigid) to duplicate roughness and then to replace the gravel with the same gradation of bottom sediment sizes in the movable bed models.

Yalin (1971) made an elaborate analysis of alluvial bed forms for hydraulic modeling. The energy slope, E_r, is taken to be:

$$E_r = \frac{Y_r}{X_r} \tag{17}$$

where Y_r is the ratio of the vertical scale and X_r is the ratio of horizontal scale. The total energy dissipation E is decomposed into three parts, E_1, E_2, and E_3 to express energy losses due to skin friction, form friction, and change of river regime, respectively.

Using a similar approach to that of the Nanjing Research Institute (1978), Yalin suggested:

$$E_1 = \frac{1}{\frac{1}{\kappa} \ell n \left(11 \frac{R}{k_s} \right)} \tag{18}$$

For alluvial bed forms friction, Yalin suggested:

$$E_2 = \frac{1}{2} \frac{\Delta^2}{\Lambda R} \tag{19}$$

when Δ and Λ are, respectively, the height and length of the bed forms. Vanoni and Hwang (1967) found that the Darcy-Weisbach friction factor, f_b'' for form roughness can be expressed by the following relationship:

$$\frac{1}{\sqrt{f_b''}} = 3.5\log\frac{R_b}{e\Delta} - 2.3 \tag{20}$$

where R_b is the hydraulic radius of the bed and e is an exposure parameter. The parameter, e, actually is the ratio between the horizontal projection of the lee faces of the ripples or dunes and the total bed area.

Wang (1983) determined from detailed direct measurements of two-dimensional dunes in the laboratory that f_b'' is a function of Δ/h and Δ/Λ.

Yalin also found that:

$$E_3 = \xi\frac{h}{L_*} \tag{21}$$

where ξ is a proportionality factor and L_* is a characteristic length. The total energy loss E_r will be:

$$E_r = \frac{E_{1m} + E_{2m} + E_{3m}}{E_{1p} + E_{2p} + E_{3p}} = \frac{Y_r}{X_r} \tag{22}$$

where the subscripts m and p denote those quantities for model and prototype, respectively. If one ignores E_2 and E_3, then:

$$E_{1m} = \frac{Y_r}{X_r}E_{1p} \tag{23}$$

and

$$\delta = E_{1m}/E_{1p} \tag{24}$$

Even if one ignores E_2 and E_3, the condition for Er is difficult to achieve in practice for all ranges of flow conditions.

Yalin further stated that in rivers, $\Delta/\Lambda < 1/10$ and $(\Delta/h) < (1/5)$. Therefore, $E_{2p} < 1/100$ in rivers and can be neglected. Since it was assumed by Yalin that E_3 for the model is always proportional to that of the prototype, and thus, the only friction factor to be considered is the skin friction. Yalin derived the following three governing criteria:

$$d_r = \frac{1}{\sqrt{\delta Y_r}} \tag{25}$$

$$d_r^3 = \frac{1}{(\gamma_s - \gamma)_r} \tag{26}$$

and

$$\sqrt{\delta} = \left[1 + \frac{\ell n\, Y_r/d_r}{\ell n\, (11 h_p/d_p)} \right] \tag{27}$$

According to Yalin,

1) for: $h/d < 1000$, $R_* < 20 \rightarrow$ Ripples
$R_* > 20 \rightarrow$ Dunes

2) for: $h/d > 1000$, $R_* < 8 \rightarrow$ Ripples
$R_* > 24 \rightarrow$ Dunes

Bogardi (1974) stated that the Shields number, $F_s = \tau/(\gamma_s - \gamma)d$ can be transformed into the following expression:

$$F_s = 458.7 \left(\frac{1}{\beta}\right)^{1.417} R_*^{-5/2} \tag{28}$$

where

$$\frac{1}{\beta} = d^{0.882} \frac{u_*^2}{gd} \tag{29}$$

Bogardi further found that β is a good indicator for the prediction of the occurrence of different bed forms and the incipient motion criteria. He stated that for β value greater than approximately 330–550, incipient motion will not start. Both the zero sediment-load criterion by Shields number less tha 0.05 and the laminar sublayer criterion can be replaced by his criterion based on β values.

Different forms of trial and error solutions are being used by various agencies to duplicate field frictional factor in the physical model. The first step is to estimate the relative importance of skin friction and form friction. Frequently, the magnitude of these frictional factors are about the same in the field condition and it is almost impossible to choose a sediment material and size that will provide similar results to both skin and form frictions in the field for all interested flow conditions. Actual experimental curves (between Q_r/X_r and S_r/X_r) obtained from data collected in the physical mode may be important to determine the proper scales. Q is the flow discharge. This will be explained later. Often onemay wish to adjust the model results to field conditions rather than choosing the exact appropriate sediment material. Of course, if bed form is a major investigation factor, the be form should be properly duplicated in the model.

2.3 Suspended Load

According to the well-known O'Brien-Rouse-Ippen equation, the vertical suspended sediment concentration can be expressed by the following expression:

$$\frac{C_y}{C_a} = \left[\frac{a}{y}\frac{(h-y)}{(D-a)}\right]^{\omega/\kappa u_*} \tag{30}$$

where C_y and C_a are respectively the suspended sediment concentration at vertical levels of y and a and ω is the terminal fall velocity of the sediment particles. For low sediment concentrations, κ is taken to be a constant of 0.4 and the most important parameter would be ω/u_*. It may be assumed tha similar suspended sediment distribution can be found by using the same ratio of ω/u_* in the laboratory as that in the field. Of course, for a distorted model, where the vertical model scale is different from that of the horizontal model scale, one must be careful in using this ω/u_* value.

Nanjing Research Institute in China (1978) also indicated the selection of the following expression to indicate suspended load capability, S_*:

$$S_* = \frac{\kappa_*}{C_o}\gamma\frac{\gamma_s}{(\gamma_s-\gamma)}\frac{V^3}{g}h\omega \tag{31}$$

After certain assumptions, the Nanjing Research Institute (1978) adopted:

$$S_{*r} = \frac{\gamma_{sr}}{(\gamma_s-\gamma)}S_r\frac{V_r}{\omega_r} \tag{32}$$

In order to achieve similarity for suspended load, the settling phenomena, the diffusion characteristics, the capability to transport and settling and erosion time must all be the same. According to the Nanjing Research Institute, the following four relationships must be valid for these four conditions, respectively:

$$\frac{V_r}{\omega_r} = \frac{X_r}{h_r} \tag{33}$$

$$V_f = V_r \tag{34}$$

where $V_f = 1.5\ln(11h/k_s)\sqrt{(\gamma_s - \gamma)/\gamma gd}$.

$$S_{*r} = \frac{\gamma_{sr}}{(\gamma_s - \gamma)_r} \tag{35}$$

where Equation 36 is derived from Equation 33 by taking $S_r V_r / \omega_r = 1$ and

$$t_{sr} = \frac{X_r}{V_r C_r} \tag{36}$$

where t_{sr} is the ratio of time for erosion and settling, and C is the sediment concentration in weight. If one requires the same time scale for both bed load motion and suspended load motion, t_{sr} (Equation 27) = t_{br} (Equation 7). This condition is difficult but possibleto attain.

3. CRITICAL ISSUES ON MOVABLE BED MODELS

The preceding sections discussed current knowledge on similarity criteria, and certain critical issues will be presented here.

3.1 Distortion Limitation

In order to examine sediment movement, the vertical scale of the physical model should be smaller than its horizontal scale. A common rule of thumb is to restrict the vertical scale to be at least one-third of the horizontal scale. This distortion certainly affects the energy slope, roughness development and scaling, time scales of both suspended load and bed load, scour and deposition, and many others. Perhaps we can discuss and share our experience in the results, as well as wisdom, of using greater distortion under certain circumstances.

3.2 Movable Bed Models to Investigate Both Suspended Load and Bed Load

There is a constant interaction between suspended load and bed load, and, on the other hand, many researchers believe that suspended load and bed load are related to different flow parameters. The possibility of conducting an experiment to investigate bed load and suspended load simultaneously is still under investigation. The time scales of these two loads may be different and there may also be an interference of bed load movement on suspended load movement. We should perhaps study the problem with the following considerations: (1) the bed load and suspended load have sediment with the same or different densities; (2) the bed load and suspended load have overlapping sediment sizes; (3) the roughness and bed forms of bed may influence suspended load movements.

3.3 Model Calibration and Verification

It is very difficult to calibrate and verify a movable bed model for a wide range of flow conditions because not all the governing model criteria can be observed. Developments of bed forms and bed roughness resistance in the field are not easily duplicated in the model for all flow conditions. Are we satisfied if roughness for a certain dominant discharge range can be duplicated in the model, and what is the dominant flow discharge range for a certain model test? Sometimes, a movable bed model was declared calibrated with a certain sediment gradation range during verification, then another sediment density with the same size gradation range was used during actual testings. Is there a better way of model calibration and verification?

3.4 Time Scales

Time scales for flow, bed load, suspended load, and scour development may not be the same as desired. Under what conditions can we allow for certain deviations? Frequently, a certain flow sequence was used to test the results of changing flow behavior. How can we take care of the different time scales and possible scale distortion for these cases?

3.5 Bed Load Criteria

The ratio betwen τ (bottom local shear stress) and τ_c (the critical local bottom shear stress for incipient motion) is a much simpler criterion to use than the modeling of absolute bed load rates. Under what conditions can we use this ratio and for what purposes?

3.6 Non-Uniform Sediment Sizes

It is rather difficult to deal with non-uniform sediment sizes for the following reasons: (1) a reliable sediment transport rate equation for non-uniform sediment sizes is not available; (2) the effect of non-uniform sediment sizes on bed form and bed roughness is not established; (3) the sediment sizes on the bed and at certain distances below the bed change with flow and time; as well as other reasons. It would be interesting to exchange our experiences on the procedure of dealing with this issue.

3.7 Scour Dimensions and Shapes

Movable bed models are frequently used to test scour problems. Often it is said that only qualitative results can be obtained by scour tests in the model. The design of scour tests should be improved so that more quantitative results can be obtained from model tests. Of course, the possible distortion of scour pattern in a distorted model test would be useful to learn.

3.8 Cohesive Sediment

Perhaps we could discuss possible model studies that can be conducted to investigate movement of cohesive material in the field.

3.9 Interaction Between Physical Models and Mathematical Models

Mathematical models have been developed rather rapidly in the past decades. It is still believed that physical hydraulic models should be coupled with mathematical models to solve for complex three dimensional flow phenomena (especially for the involvement of scour and sediment deposition under which the physical processes are not completely known). Current status and future trends should perhaps be discussed here. This will lead to my last item as stated below.

3.10 Future Research Approaches

A discussion on the best approaches to improve our knowledge will be the last main item in this workshop.

Each researcher has his own list of important items to be discussed and the purpose of this presentation was to stimulate our thoughts on this matter. It is extremely difficult to get this group of researchers together. Hopefully, we can sincerely share our knowledge on this important subject.

4. REFERENCES

Bogardi, J. 1974. *Sediment Transport in Alluvial Streams.* Water Resources Publication, Littleton, Colorado.

Chien, N. 1980 (Aug). A Comparison of the Bed Load Formulas, *Shuili Xuebao (Hydraulic Journal)*, No. 4 (in Chinese).

Einstein, H. A., and Chien, N. 1956. Similarity of Distorted River Models with Movable Beds, *Trans. ASCE*, Vol. 121, 440–457.

Nanjing Institute of Hydraulic Research. 1978. *Similarity Laws of Total Load Model with Practical Example*, Symposium On Sediment Model Tests, Beijing, China.

Vanoni, V. A., and Hwang, L. S. 1967 (May). Relation between Bed Form and Friction in Streams, *Journal of Hydraulic Division*, ASCE, HY3.

Wang, S. S. 1984. Hydraulic Models for Rivers and Hydraulic Structures, *Theory and Procedure of Model Experiments*. T. C. Zho, Editor. Beijing, China: Hydraulics and Power Press (in Chinese).

Wang, W. C. 1983. *Flow Characteristics Over Alluvial Bed Forms*. Ph.D. Thesis, Colorado State University.

Yalin, M. S. 1979. *Theory of Hydraulic Models*. New York: MacMillan.

PHYSICAL MODELLING OF SEDIMENT TRANSPORTING FLOWS

M.S. YALIN and M.S. KIBBEE
Department of Civil Engineering
Queen's University
Kingston, Ontario
Canada K7L 3N6

ABSTRACT. A method is suggested for the design of a conventional physical model of a river transporting cohesionless sediment. It is assumed that the model is Froudian and that, in general, it is distorted. A particular emphasis is given to the similarity of friction.

1. INTRODUCTION

The purpose of the present paper is to suggest a method for the determination of scales of a sediment transporting river model. It is assumed that the model is Froudian and that, in general, it is distorted: the sediment is cohesionless. The present approach to the determination of scales differs substantially from the classical approach, and this difference is explained below.

2. METHODS OF DETERMINATION SCALES

According to the classical approach in Langhaar, 1962; Sedov, 1960; and Yalin, 1987, the dynamic similarity of a phenomenon is *achieved* if the model and prototype equality of the dimensionless variables of that phenomenon is *provided*. Indeed, since any dimensionless property π_j of a phenomenon is but a *different* function (f_j) of the *same* N dimensionless variables X_i ($i = 1, 2, \dots N$), i.e. since

$$\pi_j = f_j(X_1, X_2, \dots X_N), \tag{1}$$

H. W. Shen (ed.), Movable Bed Physical Models, 13–22.

The dynamic similarity, implied by

$$\pi_j'' = \pi_j', \textit{ that is by } \lambda_{\pi_j} = 1, \tag{2}$$

can certainly be achieved (for any π_j), if

$$X_i'' = X_i' \textit{ that is } \lambda_{X_i} = 1 \tag{3}$$

is provided for all N dimensionless variables X_i. The great advantage of this method lies in the fact that it supplies the scale relations (from Equation 3) without "knowing" the mathematical form of the functions f_j, and without "knowing" the prototype values X_i'. On the other hand, as is well known, often one cannot fulfill Equation 3 for all the dimensionless variables (when $\lambda_g = \lambda_\rho = \lambda_\nu = 1$ is imposed) and in such cases the effectiveness of the method reduces considerably. Indeed, in such cases some variables (labelled as "unimportant") are ignored and Equation 3 is fulfilled only for a subset of the set X_i ($i = 1,2, \ldots N$). Needless to add that the judgement on "unimportance" is very subjective, and the fact remains that the rejected variables seriously undermine the model tests (because of their inequality in model and prototype) in the form of unpredictable "scale effects".

Suppose now, that the model is distorted ($\lambda_y \neq \lambda_x$). In this (widely used, and yet seldom justified) case, the scales of the dimensionless properties π_j are, in general *different from unity*: for example the scale of the slope itself is $\lambda_s = \lambda_y/\lambda_x \neq 1$, the scale of the dimensionless friction factor is $\lambda_c = (\lambda_s)^{-0.5} \neq 1$, ... etc. Leaving aside the issue that relations such as

$$\lambda_{\pi_j} = \lambda_s^m \neq 1 \tag{4}$$

contradict the very definition of the dynamic similarity (embodied by Equation 2), it is interesting to point out that the requirements such as Equation 4 cannot be achieved by equating the dimensionless variables X_i in model and prototype (as implied by Equation 3). Indeed, consider Equation 1. The form of the function f_j is the same (in model and prototype) for any specified property π_j. If in addition to this, the values for all X_i are also identical ($X_i'' = X_i'$), then the values of π_j must necessarily be identical too. But $\pi_j'' = \pi_j'$ i.e.

$$\lambda_{\pi_j} = 1$$

contradicts the requirement of Equation 4 and therefore, the (classical) attempt to provide the model and prototype equality of the dimensionless variables loses its meaning if the model is distorted. In other words, in the case of a distorted model ($\lambda_y \neq \lambda_x$), the requirements such as Equation 4 can be satisfied only if (at least some of) the dimensionless variables X_i are <u>not</u> identical in model and prototype.

For a dimensionless property π_j of a distorted model, we have

$$\pi_j' = f_j(X_1', X_2', ... X_N'), \tag{5}$$

and

$$\pi_j'' = f_j(X_1'', X_2'', ... X_N'') = f_j(\lambda_1 X_1', \lambda_2 X_2', ... \lambda_N X_N') \tag{6}$$

which yield (considering Equation 4)

$$\frac{\pi_j''}{\pi_j'} = \lambda_{\pi_j} = \lambda_s^m = \frac{f_j(\lambda_1 X_1', \lambda_2 X_2', ... \lambda_N X_N')}{f_j(X_1', X_2', ... X_N')} \tag{7}$$

Consider the case where the form of the functions f_j are known for k – number of properties π_j. In this case we have k – number of (known) Equation 7 which can be shown symbolically as

$$\lambda_s = \Phi_j(\lambda_{X_i}, X_i'), \qquad (i = 1,2,...N, \; j = 1,2,...k) \tag{8}$$

Clearly, the determination of scales λ_{X_i} from a set of k equations such as Equation 8 will (in contrast to the classical method reflected by Equation 3) require a knowledge of the form of the form of the functions ϕ_j and on the prototype values X_i' (which is not a severe limitation, as one usually knows the numerical order of the characteristics of the prototype one intends to model).

Clearly, the number k of the relations such as Equation 8 should not exceed the number n of the (unknown) scales λ_{X_i} for if n = k, then one cannot choose even one scale freely, and if k > n then the system of k equations will become indeterminate.

3. MODELLING OF SEDIMENT TRANSPORT

The mass transport of sediment is specified (at a given location and instant) by six characteristic parameters, which can be considered to be

$$\rho,\ \nu,\ h,\ S,\ \gamma_s,\ D \tag{9}$$

Since the distorted Froudian model under consideration operates with the prototype fluid (water), i.e. since $\lambda\rho = \lambda\nu = 1$, the model is determined by the following n = 4 scales

$$\lambda_h, \lambda_s, \lambda_{\gamma_s}, \lambda_D \tag{10}$$

Knowing the prototype values of characteristic parameters, Equation 9, one can compute the prototype values of N = 6 – 3 = 3 dimensionless variables $X_1 = X$, $X_2 = Y$ and $X_3 = Z$ of the two phase motion "en mass" from Equations 1a, 2a and 3a in Appendix A. Similarly, one can compute the prototype values of the dimensionless characteristics ξ, Y_{cr} and σ (from Equations 4a, 5a and 6a respectively). Equation 5a is the analytical expression of the tranport inception curve $Y_{cr} = \phi(\xi)$. [The graph of Equation 5a passes right through the midst of the points forming the experimental curve $Y_{cr} = \phi(\xi)$]. In fact, exactly the same can be said with regards to the Equations 7a, 12a and 13a. Their graphs, too, pass through the midst of the point-patterns forming the experimental curves of B, $(\Delta/\Lambda)_d$ and $(\Delta/\Lambda)_r$. The analytic expressions are introduced herein in order to avoid the utilization of graphs and to obtain the results only with the aid of computation. See Equation 5.

The model values X_i'' of the dimensionless variables, are determined as the products of the prototype values X_i' with the corresponding combinations of scales. e.g. Y'' is given by

$$Y'' = \left(\frac{\lambda_\rho \lambda_{v_*}^2}{\lambda_{\gamma_s}\lambda_D}\right) Y' = \left(\frac{\lambda_h \lambda_s}{\lambda_{\gamma_s}\lambda_D}\right) \tag{11}$$

At this stage it should be pointed out that out of four scales in Equation 10 it is only three, namely λ_s, λ_{γ_s} and λ_D, that are really unknown, the scale λ_h can be regarded as known (for it is virtually determined by h'). Indeed, the prototype flow depth h' can be 2 m and it can be 20 m (Mississippi, Amazon, etc.). Yet the flow depth h'' in a conventional model is always between 15 cm and 35 cm, for example. Thus, if one aims at a model having e.g. $h'' = 25$ cm, then

$$\lambda_h = \frac{0.25}{h'} \tag{12}$$

Hence, it is in fact only n = 3 scales (namely λ_s, λ_{γ_s} and λ_D) which are unknown in the true sense of the word, and in accordance with method reflected by Equation 8 they must be determined with the aid of k = 2 pertinent relations (k < n). The first relation will be formed with the aid of the relative mobility number Y/Y_{cr}. Its equality in the model and the prototype yields.

$$\lambda_Y = \lambda_{Y_{cr}} \neq 1 \quad i.e. \quad \lambda_s = \lambda_{Y_{cr}} \lambda_{\gamma_s} \lambda_D \lambda_h^{-1} \tag{13}$$

The second relation will be formed by using the dimensionless Chezy friction factor

$$c = \frac{v}{v_*} = \frac{v}{\sqrt{gh}}\frac{1}{\sqrt{S}} = \frac{Fr}{\sqrt{S}} \tag{14}$$

This relation indicates that if the distorted model ($\lambda_s \neq 1$) is Froudian ($\lambda_{Fr} = 1$) then

$$\lambda_c = \frac{1}{\sqrt{\lambda_s}} \quad (\neq 1). \tag{15}$$

On the other hand, since the (scalar) energy is additive, the friction factor c, or to be more precise, the square of its reciprocal ($1/c^2$) can be given by the sum

$$\frac{1}{c^2} = A_f + A_d + A_r \tag{16}$$

where A_f, A_d, and A_r are the energy losses due to pure friction, due to dunes and due to ripples (superimposed on dunes) respectively:

$$A_f = \left[2.5 \ln\left(b\frac{Z}{2}\right)\right]^{-2} \tag{17}$$

$$A_d = \frac{1}{2}\left(\frac{\Delta}{\Lambda}\right)_d^2 2\pi \tag{18}$$

$$A_r = \frac{1}{2}\left(\frac{\Delta}{\Lambda}\right)_r^2 \frac{1000}{Z} \tag{19}$$

(Here b can be computed from (7a) and (8a) in Appendix A, the reason for $Z/2 = h/(2D)$ on the right of Equation 18 is because the size k_s of the granular "skin roughness" can be approximated by $k_s \approx 2D$ (Kamphuis, 1974), (Yalin, 1977). The factor 2π and $1000/Z$ approximate Λ_d/h and Λ_r/h respectively. The values of the dune steepness $(\Delta/\Lambda)_r$ and the ripple steepness $(\Delta/\Lambda)_r$ can be computed from the expressions 10a to 13a in Appendix A. A more extensive explanation of the characteristics involved in Equations 18 to 20 can be found in Yalin and MacDonald (1987).

Now, from Equations 15 and 16 it follows that

$$\lambda_s = \frac{A_f'' + A_d'' + A_r''}{A_f' + A_d' + A_r'} \tag{20}$$

must be valid (to ensure the similarity of the flow profile). Observe that the equality $A''_i = A'_i$ for all three components of the energy loss, would yield unity for the right hand side of Equation 20, which would be inconsistent with the requirement $\lambda_s \neq 1$. Hence the energy losses must necessarily be distorted: the numerator of the right hand side of Equation 20 must be λ_s times larger than the denominator.

It follows that the model i determined by such λ_s, λ_D, and λ_{γ_s} which satisfy, simultaneously, Equations 13 and 20. It is true, that the system Equations 13 and 20 has an infinite number of solutions with respect to λ_s, λ_D, and λ_{γ}s. However, only very few of these (mathematical) solutions can be used to design a practical model. Indeed, a practical model can be realized only for that (small) subset of the solutions λ_s, λ_D, and λ_{γ_s} which satisfy the following (rather severe) conditions, i.e. which "fall" into the three dimensional "box":

$$0.03 \le \lambda_{\gamma_s} \le 1 \tag{21}$$

$$1 \le \lambda_s \le \lambda_h^{-1/2} \tag{22}$$

$$\lambda_h \le \lambda_D \le 3.22 \tag{23}$$

The condition Equation 21 implies that the model bed material should not be heavier than sand and no lighter than polystyrene (to ensure it sinking). The left side limit of Equation 22 reflects the undistorted model, the right hand side signifies the "Wallingford experience" $\lambda_y = \lambda_x^{2/3}$ (see Yalin, 1971). The relation Equation 23 implies that the grain size scale should not be smaller than the vertical model scale, and that it should not be larger than that which follows from the classical approach $\lambda_X = 1$; $\lambda_Y = 1$ (i.e. from $\lambda_{\gamma_s}\lambda_D^3 = 1$) where $\lambda_{\gamma_s} = 0.03$.

4. COMPUTATION PROCEDURE

The present approach is computer oriented. Knowing the prototype values h', S', and D' (and taking into account that $\rho' = 10^3$ kg/m^3, $\nu' = 10^{-6}$ m^2/s, $\gamma_s' = 1.65\,\gamma'$) one computes all the pertinent prototype characteristics from (1a) to (14a) in Appendix A. Next, one specifies λ_h (with the aid of Equation 12) and **adopts** a value for λ_{γ_s} from the range Equation 21. Thus only two scales, namely λ_s and λ_D, must be determined. This can be achieved by solving (numerically) the system of two Equations 13 and 20. The denominator of Equation 20 is known – it is the prototype characteristic $[c']^{-2}$ (already computed from (14a)). The model values in the numerator of Equation 20 are also determined by the relations (1a) to (14a) in Appendix A. However, this time, these relations must be evaluated for model characteristics and the corresponding combination of scales, e.g. as

$$h'' = \lambda_h h', \quad S'' = \lambda_s S', \quad Y'' = \left(\frac{\lambda_s \lambda_h}{\lambda_D \lambda_{\gamma_s}}\right) Y', \quad Z'' = (\lambda_h \lambda_D^{-1}) Z' \ \dots \ etc.$$

Hence for a given prototype, and the adopted λ_h and λ_{Y_s}, the numerator on the right side of Equation 20 will contain only two unknowns (λ_s and λ_D). If the computation procedure is by iteration, it may be helpful to determine the "first approximation" $(\lambda s)_1$ by solving the "approximate set", as shown in the following Equation 24 and 25.

$$(\lambda_s)_1 \sim \left[\lambda_{Y_s}\lambda_h^{-1}\right](\lambda_D)_1 \tag{24}$$

$$(\lambda_s)_1 \sim c'^2 \left[[2.5 \ln(5.5\ \lambda_h(\lambda_D)_1^{-1}\ Z')]^{-2} + A'_d + \frac{A'_r}{Z'} \right] \tag{25}$$

Where A_d' and A_r' are simply the prototype values of A_d and A_r (given by Equation 18 and Equation 19 and, of course, also by (12a) and (13a)).

In the same manner, one can determind λ_s and λ_D for another adopted value of λ_{Y_s} (from the range Equation 21) and so on. Thus, one arrives at a set of the theoretically possible "trios" λ_{Y_s}, λ_s, λ_D (all of which correspond to the specified λ_h). Examining this set in light of the "practical constraints" Equations 22 and 23 one reveals the most suitable trio.

The examples of scale computations using the present method are shown for four different prototypes in Appendix B. Here the last column (namely, w ratio) is the value of the ratio λ_w/λ_{v_*} (the closer this ratio is to unity the more the likelihood for an adequate reproduction of the distribution of suspended load). The subset of solutions satisfying the constraints Equations 21, 22 and 23, and thus which can be regarded as acceptable are indicated.

5. ADDITIONAL REMARKS

i) Why the present method rests on

$$\lambda_Y = \lambda_{Y_{cr}} \quad \text{(Equation 13)}$$

and not the customary

$$\lambda_Y = 1 \quad ? \ldots$$

The answer to this question lies in the fact that in the present approach the model and the prototype values of X or of ξ are (necessarily) not equal. But if $X_{cr}'' \neq X_{cr}'$ (or $\xi'' \neq \xi'$) then in general, $Y_{cr}'' \neq Y_{cr}'$. Suppose now, that Y numbers are made equal in model and prototype ($Y'' = Y'$). In this case the ratio $Y/Y_{cr} = \tau_o/(\tau_o)_{cr}$, implying the **stage** of sediment transport will not be equal in model and prototype. For example, if this non-equality is such that

$$\frac{Y}{Y''_{cr}} < 1 \qquad \textit{while} \qquad \frac{Y}{Y'_{cr}} > 1$$

then in the prototype the transport will be present, whereas in the model it will be absent (for the same Y) which is obviously unacceptable. The same dissimilarity will be present with regard to any other aspect determined by the stage $\tau_o/(\tau_o)_{cr}$ (initiation or disappearance of sand waves, initiation of suspended load, etc). The introduction of Equation 13 (instead of $\lambda_Y = 1$) perhaps is the most significant difference between the present method and that suggested in Yalin and MacDonald (1987).

ii) If the model design method is sound, then the scale of any property should be predictable beforehand. The following two examples should be sufficient to illustrate this point.

a) **The bed load rate** q_s can be given e.g. by the Bagnold form

$$q_s \sim v\left[\tau_o\left(\frac{c}{\bar{c}}\right)^2 - (\tau_o)_{cr}\right] \tag{26}$$

where $c/\bar{c}$ converts the total shear τ_o to the pure friction shear (if sand and waves are present). Considering that the model is Froudian ($\lambda_v = \lambda h^{1/2}$) and that $1/\bar{c}^2 = A_f$; one obtains from Equation 26

$$\lambda_{q_s} = \sqrt{\lambda_h}\,\lambda_{Y_{cr}}\,\frac{\lambda_{A_f}\lambda_c' K' - 1}{K' - 1} \tag{27}$$

where K′ stands for the (known) prototype value of the combination

$$K = \frac{\tau_o}{(\tau_o)_{cr}} A_f c^2 \tag{28}$$

b) **The time of formation** T of an erosion or deposition process can be given (per unit flow width) by

$$\int_0^T q_s\,dt = \gamma_s V_T \tag{29}$$

where V_T is the volumn of sediment eroded (or deposited) per unit flow width during T. The value of V_T can be given by

$$V_T = LH \tag{30}$$

where L is the base-length (along x of the erosion (or deposition) region, and H is its space-average depth (or height). From Equations 29 and 30 it follows immediately that

$$\lambda_T = \frac{\lambda_{\gamma_s}\lambda_x\lambda_y}{\lambda_{q_s}} = \frac{\lambda_{\gamma_s}\lambda_s^{-1}\lambda_h^2}{\lambda_{q_s}} \tag{31}$$

where λ_{q_s} is given by Equation 27. The same result can be obtained also from the Exner–Polya equation

$$\frac{\delta q_s}{\delta x} + \gamma_s \frac{\delta Z}{\delta t} = 0 \qquad (since\ L \sim x,\ T \sim t\ and\ H \sim Z) \tag{32}$$

[Suppose that Q′ = f(t′) varies considerably during T′. In order to achieve the similarity of flow action on the model bed, Q″ must vary during T″ in an analogous manner. But this means that the time scale of the time–variable flow is also gven by the right hand side of Equation 31].

6. REFERENCES

Kamphuis, J. W.: (1974) 'Determination of Sand Roughness for Fixed Beds,' *Journal of Hydraulic Research*, **12**, pp. 2.

Langhar, H. L.: (1962) *Dimensional Analysis and Theory of Models*, John Wiley, New York.

Sedov, L. I.: (1960) *Similarity and Dimensionless Methods in Mechanics*, Academic Press Inc., New York.

Yalin, M. S.: (1971) *Theory of Hydraulic Models*, Macmillan, London.

Yalin, M. S. and N. J. MacDonald: (1987) *Determination of a Physical River Model with a Mobile Bed*, XXII Congress IAHR, Lausanne.

Yalin, M. S.: (1977) *Mechanics of Sediment Transport*, (Second Edition), Permagon Press, Oxford.

LIST OF SYMBOLS

f	gravity acceleration	$Fr = \frac{v^2}{gh}$	flow Froude number
ρ	fluid density	X	grain Reynolds number
ν	kinematic viscosity	Y	mobility number
h	flow depth	Z	dimensionless flow depth
v	mean flow velocity	ξ	$= X_{2/3}\ Y^{-1/3}$ material number
v_*	shear velocity	κ	= 0.4 Von Karman constant
D	typical grain size	$Y_{cr} = \phi(\xi)$	critical value of Y

γ_s specific grain weight in fluid

ks sand wave roughness

Δ sand wave height

Λ sand wave length

$\sigma = [(Y/Y_{cr})-1]$ relative flow intensity

$R_* = v_* k_s/\nu$ roughness Reynolds number

$B = f(R_*)$ roughness function

c dimensionless Chezy friction factor

If Q is any quantity, then Q′ and Q″ are prototype and model values of Q: $\lambda_Q = Q''/Q'$ is the scale of Q. λ_y and λ_x are vertical and horizontal model scales respectively.

Subscripts r and d signify ripples and dunes respectively.

DIMENSIONAL ANALYSIS, DYNAMIC SIMILARITY, PROCESS FUNCTIONS, EMPIRICAL EQUATIONS AND EXPERIENCE - HOW USEFUL ARE THEY?

PETER ACKERS
Hydraulics Consultant
4 Glebe Close
Moulsford, Wallingford, OX10 9JA
United Kingdom

ABSTRACT. This paper points out the limitations of dimensional analysis in scale selection for mobile bed models, and compares the roles of dynamic similarity, the regime approach and sediment transport and alluvial resistance functions, in modelling sand bed rivers.

1. DIMENSIONAL ANALYSIS OR DYNAMICS SIMILARITY?

Only in very simple cases where the number of variables in the system is limited and precisely definable does dimensional analysis *by itself* provide the basis for scale selection and model operation. Modellers can seldom base their scales on rigorous dimensional analysis even if they claim to do so: in many cases the issue is compromised or obscured by implicit introduction of prior knowledge based on similarity or of the physical process. When one considers a process as complex as sediment transport, the deficiencies of dimensional analysis unsupported by similarity arguments or empirical knowledge are apparent.

The minimum set of basic quantities which influence the process of sediment transport in two-dimensional, free-surface flow are the unit mass of fluid, ρ, the unit mass of solids, ρ_s, the viscosity of the fluid, μ, particle diameter, D, water depth, d, shear velocity at the bed N(gds) denoted v_*, and acceleration due to gravity, g. Dimensional analysis yields four groups:

H. W. Shen (ed.), Movable Bed Physical Models, 23–30.

$$Re_* = \rho v_* D/\nu \tag{1}$$

$$Y = v_*^2/(s-1)gD \tag{2}$$

$$Z = d/D \tag{3}$$

$$s = \rho_s/\rho \tag{4}$$

Using Einstein's (1950) non-dimensional expression for sediment transport:

$$\phi = q_s/\rho[s-1)gD]^{3/2} \tag{5}$$

where q_s is the sediment transport rate as submerged weight per unit time per unit width, it follows that:

$$\phi = \text{function}[Re_*;Y;Z;s] \tag{6}$$

The mobile bed friction factor must also depend on these same parameters. Most transport theories use the above parameters or their equivalent. For example, Ackers and White (1973) replaced the above particle Reynolds number, Re_*, by:

$$D_{gr} = D[g(s-1)/\nu^2]^{1/3} \tag{7}$$

where $\nu = \mu/\rho$

If one is to use dimensional analysis as the sole basis for scale selection, there is no choice but to achieve equality of all these dimensionless groups. Departure from this principle is only possible if one can prove that one or more of them has no influence. However, only a 1:1 scale is capable of satisfying all four groups with water as the fluid. Compromise is thus essential, especially when one takes account of the usual requirement for Froude similarity if the water motion itself is dependent on gravity.

The physical significance of each of the dimensionless groups is important if one is to proceed further. D_{gr} expresses the relative influence of immersed weight and viscosity on the motion of individual particles; Y is the ratio of the bed shear stress to the weight of a grain thick layer of sediment; Z is the geometric effect of flow depth in relation to particle diameter and is particularly relevant to resistance; s brings in the possible influence of particle inertia, as opposed to weight which is accounted for in Y.

Most sediment transport researchers would agree that the group denoted by s has no detectable effect, the immersed weight being appropriately accounted for by Re_* (or D_{dg}) and Y. Yalin (1971) demonstrates that even omitting the fourth group no practical set of scales emerges. From Z:

$$_sD = {_sd} \tag{8}$$

Hence from Re*

$$_sv_* = 1/{_sd} \tag{9}$$

and hence

$$_sS = 1/{_sd}^3 \tag{10}$$

where prefix s denotes "scale of" and S is the channel gradient. Clearly such extreme exaggeration of slope is impractical, and would not in any case be compatible with Froudian simulation of flow. The neglect of Z would be an over-simplification, although some transport formulae such as Engelund and Hansen (1987) omit that group. If only the first two groups are regarded as significant, then there is greater flexibility in scale selection:

$$_s(s-1) = 1/{_sD}^3 \tag{11}$$

which provides equality of D_{gr}, and

$$_sS = {_sD}^{-2}\,{_sd}^{-1} \tag{12}$$

where the slope scale is usually taken as the vertical exaggeration of the model. Implicit in this is the assumption that the alluvial resistance is also determined by Re* and Y only. This requires the use of light weight material in very large models in those situations where the sediment is not coarse enough to warrant neglect of particle Reynolds number.

Yet many apparently successful models have been made of sand bed rivers and estuaries which did not follow those rules.

2. REGIME FORMULAE

The regime approach to modelling is essentially empirical and may be derived from the "classic" regime equations (Inglis, 1949):

$$\text{width,}\quad W = C_1 Q^{1/2} \tag{13}$$

$$\text{depth,} \quad d - C_2Q^{1/3} \tag{14}$$

$$\text{slope,} \quad S - C_3Q^{-1/6} \tag{15}$$

The three coefficient values depend on sediment size and on sediment transport expressed as a concentration, but the corollary is that if sediment size is the same in model and prototype, simulation of sediment concentration (and hence equivalent sediment time scale to water movement time scale) is achieved with a vertical exaggeration of the inverse square root of the vertical scale in a Froudian model, when:

$$_sW - _sx - _sd^{1-1/2} - _sV^3 - _sQ^{1/2} - _sS^{-3} \tag{16}$$

where x is the length dimension of the system.

Laboratories which have perhaps implicitly used regime methods for scale selection are likely to have had much experience with them and so their scale selection is moderated by the results of that experience. A sand bed river model that the author was responsible for many years ago was based on the interpretation of regime experiments at laboratory scale, but the model tended to generate depths that were low and velocities correspondingly high. Adjustment was made to the scales and the model was operated at discharges and velocities some 20% above Froude scale. This type of experience has been built into the activities of many laboratories, of course.

4. DYNAMIC SIMILARITY OR PROCESS FUNCTIONS?

The term "process function" is used to describe the formulae that are available to represent some physical process. The conventional friction diagram based on the Colebrook – White equation provides a basis for designing models where friction over rigid boundaries is significant: this is an example of the use of a process function to aid modelling.

Methods for the selection of scales for mobile bed models on the basis of preferred sediment transport functions will be found in the literature. Frequently these are a restricted form of applying dimensional analysis or similarity arguments in that the dimensionless groups appearing in the preferred function will be taken as a sufficient set. This is a logical approach and as several of the well known functions depend on similar reduced sets of non-dimensional parameters, there is some agreement about the modelling laws that may be derived. The fewer parameters that appear in the transport function chosen, the greater the flexibility in scaling, but choosing the transport formula on the basis of ease of manipulation rather than its proven record of performance at both field scale and laboratory scale might not lead to good simulation!

Let us consider one of the more recent total load functions as a basis of scale selection. The author's choice of the function (Ackers and White, 1973) is based on familiarity with its application and performance; it is introduced here as an example of the most flexible use of process equations in guiding scale selection, but other recent transport functions of comparable reliability may be used in a similar way. The same parameters as were used in this transport formula were also used by White, Paris and Bettess (1980) in deriving their resistance function for alluvial streams in the lower regime of bed features, and this provides the convenience of simultaneous solution of the resistance and transport functions (see for example, Ackers, 1983).

The example chosen is of a major sand bed river, D = 0.2 mm, operating at discharge intensities from 10 to 40 $m^3/s/m$ width, with a sediment rating curve for total bed material transport ranging from a "charge" of 2000 mg/l to 8000 mg/l over that flow range. Joint solution of the transport and resistance functions provides the "regime" depths, velocities and gradients. The same equations may then be used to find model conditions that will simulate those depths, velocities and discharge intensities on a Froudian basis, for an assumed vertical scale. The vertical scale selected is 1/30, hence velocity scale 1/5.48 and discharge intensity scale 1/164.3. It is assumed that 0.2 mm sand will be used in the model, thereby following the practice of some traditional laboratories using regime methods whilst also ensuring that Re_* (and D_{gr}) are the same in model and prototype. Figure 1 illustrates both the full scale solution and the model scale solution. The simulation of large sand bed rivers by sand bed models is confirmed by application of recent process equations.

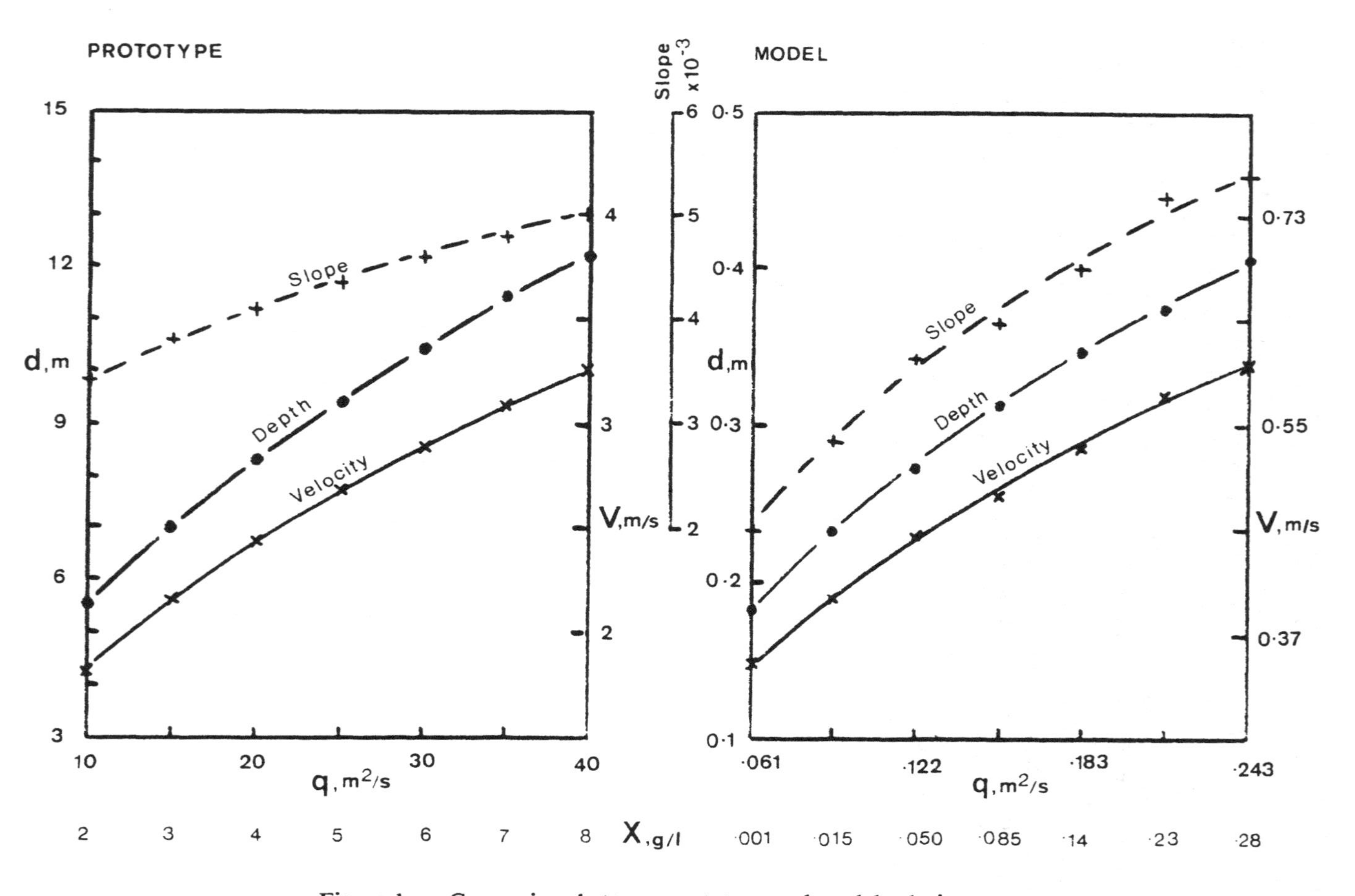

Figure 1. Comparison between prototype and model solutions.

The unknowns determined by the model solution are slope and sediment charge. This sediment supply would have to be fed into the model to achieve depths and velocities to the correct scale, given the Froudian discharge intensity scale. Slope would be self adjusting. Model sediment concentrations are considerably below the prototype values, and so the morphological time scale of this example is slower than the Froudian time scale for water movements. This slowing down is by a factor of over 20 (the ratio of X values at high flows) but this may not be of great significance provided there is sufficient time in both prototype and model for a quasi-steady regime to be achieved.

Note also that the slopes in the model turn out to be very similar to those in the prototype: marginally steeper at the highest flows. These are the head losses arising from alluvial friction over lower regime beds, excluding any additional losses arising from complex geometry of a real river. However, the approximate equality of alluvial energy gradient in model and prototype suggests that this sand bed model is most appropriate at natural scale because in theory there is no general exaggeration of the slope. This is the only restriction on the choice of a horizontal scale. In theory there is a free choice of horizontal scale, provided the system modelled can be represented with a slope scale differing from that given by the two geometric scales. In many studies of river training or the performance of a barrage and its canal intakes, a relatively short reach of river may be modelled and such models may therefore not be very sensitive to distortion of the slope scale.

It has been presumed in the above that both model and prototype operate with lower regime bed forms, but this has to be checked. Bettess and Wang Shiqiang (1986) have developed a criterion for the transition towards upper regime bed forms using the parameter:

$$UE = VS/[(g)^{1/3}D_{gr}] \qquad (17)$$

They suggest that when UE > 0.011 (calculated for the lower regime form), transition towards upper regime begins. In the example given above, the value of UE in the prototype reaches 0.011 when the discharge intensity reaches 25 $m^3/s/m$, and so at higher discharges the dunes of the lower regime would be partly washed out. Model values of UE remain below 0.011. It seems possible that in times of flood many of the major sand bed rivers of the world could be in that uncertain transition between lower regime and upper regime, whereas their models (if they have sand beds) would remain in lower regime. Thus simulation must depend very much on the observed hydraulics of such rivers, rather than being based solely on calculation. This is consistent, of course, with the wise procedures of calibration and validation, and it demonstrates also that the simulation of mobile bed channels is very dependent on alluvial channel resistance, a point not always appreciated in theoretical methods.

5. SUMMARY

- The modelling problems of interest to-day are those concerning complex flow phenomena that are not amenable to simple solution on the basis of dimensional analysis or similarity principles.

- An understanding of the phenomena themselves, expressed as process equations, is increasingly used to guide modelling.
- There is general recognition that many aspects of complex hydraulic systems cannot be satisfactorily modelled. These aspects have to be dealt with by other methods but a combination of those methods with a scale model is often the best way to a solution.
- The reliability of scale models can not be judged by theoretical exercises or from laboratory scale tests alone. Well documented measurements of prototype behaviour are the only proof of success.

6. REFERENCES

Ackers, P. and White, W.R. (1973) 'Sediment transport: New approach and analysis,' *Proc. American SCE*, Nov. 1973, **99**, HY11, 2041-2060.

Ackers, P. (1983) 'Sediment transport problems in irrigation systems design, developments in hydraulic engineering - I.' ed. Novak P. *Applied Science Publishers*, England, 151-195.

Bettess, R. and Wang Shiqiang (1986) 'The frictional characteristics of alluvial streams in the lower and upper regime,' *Hydraulics Research*, Wallingford, England, awaiting publication by Instn. Civ. Engrs., London.

Einstein, H.A. (1950) 'The bed load for sediment transportation in open channel flow,' *Ser. Tech. Bull. U.S. Dept. Agriculture & Soil Conservation*, 1026.

Engelund, F. and Hansen, E. (1967) 'A monograph on sediment transport in alluvial streams,' *Teknisk Vorlag*, Copenhagen, Sweden.

Inglis, C.C. (1949) 'The behaviour and control of rivers and canals (with the aid of models),' *Central Waterpower Irrigation & Navigation Research Station*, Poona, India, Research Pub. No. 13.

White, W.R., Paris, E., and Bettess, R. (1980) 'The frictional characteristics of alluvial streams: A new approach,' *Proc. Instn. Civ. Engrs.*, London, **69**, Part 2, Sept. 1980, 737-750.

Yalin, M.S. (1971) *Theory of hydraulic models*, Macmillan, London.

DISTORTED MODEL AND TIME SCALE EVALUATION OF MULTISCALE SUBJECTED FLUVIAL PROCESSES

TETSURO TSUJIMOTO
Kanazawa University
Department of Civil Engineering
2-40-20 Kodatsuno
Kanazawa, Japan

ABSTRACT. Fluvial process is often subjected to several different factors and thus a physical model test, which satisfies all requisite similarity criteria, is difficult to plan. Sometimes a distorted model is required and the detailed arrangements for the requisite conditions are particularly difficult to meet. With a physically reasonable analytical model, we could skillfully predict prototype behavior from partially imperfect distorted model. By taking an example in predicting scour depth around a bridge, this study demonstrated how an analytical model can be used effectively to complement a physical model study.

1. INTRODUCTION

Fluvial processes are often governed by multi-factors. For example, local scour around a bridge pier often involves several scales, as shown schematically in Figure 1. The arguments here are limited only to the spatial scales and the governing characteristic scales are listed below.

1) Sand bed materials are sometimes composed of graded materials, or sand mixtures, and thus the sediment gradation curve becomes important;

2) Flow depth and pier diameter are especially important to the local scour;

3) Dunes and bars, or small-scale and meso-scale bed forms often affect scouring process. Particularly the migration of bed forms exerts a great influence on scour;

4) A supplementary structure may be added upstream from a bridge pier for prevention of scouring or reducing scour depth. Particularly in rivers under severe degradation, such a protection work becomes remarkable, or the foundation of the pier itself often becomes severely exposed to the flow. These types of supplementaries are also subject to the scouring proces.

H. W. Shen (ed.), Movable Bed Physical Models, 31–48.

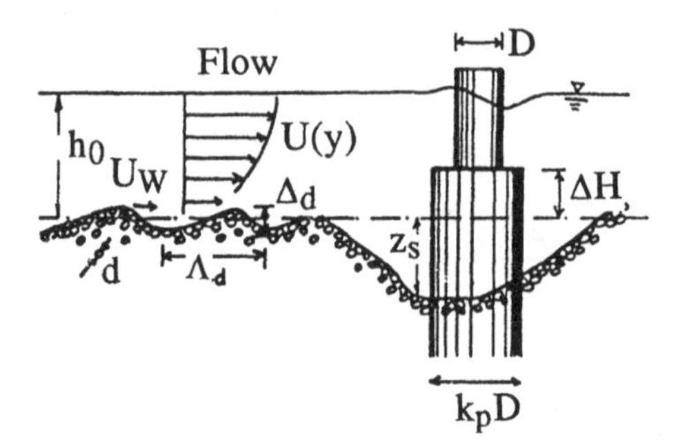

Figure 1. Schematic illustration of multiscale subjected local scour.

For fluvial hydraulic model tests, there are three scales: sediment size; the vertical size of flow; and the horizontal size of the area for the model test. These are quite independent of each other, and are much different from each other.

Generally, sediment size governs bed roughness and sediment motion, and sufficient vertical flow size is required to provide turbulent levels. On the other hand, the available space of a laboratory determines the horizontal scale of the model. The other scales, sediment size and flow depth in the model, thus often inevitably become too small. In such cases, we have a "distorted model" where several geometric scales are distorted for the reasonable sense. Reasonable sense here means that turbulent flow must exist in the model and the common pattern of sediment motion between prototype and model must be similar.

In spite of using a distorted model, some subsidiary effects on local scour as a main phenomenon cannot be realized or reproduced in the model. Moreover, for the sake of simplicity or easiness in making a physical model in the laboratory, subsidiary effects are sometimes neglected but can be considered separately by an analytical approach. In order to predict these subsidiary effects analytically, we have to derive a reasonable mathematical description of the essential mechanism and then we are able to use or extend the result of the model test to evaluate the prototype phenomenon. In other words, we have to manipulate these two approaches complimentarily.

The subsidiary effects on local scour around bridge pier are listed for examples which are in general difficult to reproduce in a model (particularly in a small and distorted model):

1) Effect of the additional scales due to the supplementary structures;
2) Effect of migration of dunes; and
3) Effect of sand mixtures.

The above three effects will be explained in Section 4, after discussion on theoretical background for the physical fluvial model test.

2. THEORETICAL BACKGROUND OF DISTORTED MODEL

An approach using a distorted model is discussed in the framework of the simplest situation: two-dimensional phenomena. If the phenomena is two-dimensional in the vertical plane, the governing non-dimensional parameters are:

$$X = u^* \frac{d}{\nu} \quad Y = \frac{u^*}{\left[\left(\frac{\sigma}{\rho - 1}\right) gd\right]} \quad Z = \frac{h}{d} \quad W = \frac{\sigma}{\rho} \tag{1}$$

in which u^* = shear velocity of flow; d = sand diameter; ν = kinematic viscosity; σ/ρ = relative density of sand; h = flow depth; and g = gravity acceleration. X is mainly concerned with the pattern of sediment motion; Y with sediment transport rate; Z with flow resistance; and the immediate effect of W is considered to be negligible.

At first, the similarity of flow dynamics is accompished by the Froude number similarity law, and according to it, the following scale relation is obtained:

$$U_r = h_r^{1/2} \tag{2}$$

in which U = flow velocity; and the subscript r means the scale (ratio) of the model to the prototype.

Next, we have to consider the similarity of sediment transport rate and it is generally achieved by the similarity law for Y value, or the well-known Shields parameter. Namely, the following equation is valid:

$$\frac{U_r^2}{\left[\left(\frac{C}{\sqrt{g}}\right)_r \left(\frac{\sigma}{\rho - 1}\right)_r d_r\right]} \tag{3}$$

in which C = Chézy coefficient. If the Manning-Strickler equation is conveniently applied here, we obtain the following equation, which gives the relation between the vertical flow rate and sediment size scale.

$$h_r^{2/3} = \left(\frac{\sigma}{\rho - 1}\right)_r d_r^{2/3} \tag{4}$$

Moreover, in order to keep the validity of the transport relation represented by Y value similarity, one should also try to produce the similarity of the pattern of sediment motion between prototype and model by X value similarity law. This relation is not so limiting and allows some range of X_r value for the given X_p value.

Roughly saying, if X_p is larger than 70, the flow is rough turbulent and sediment is transported mainly as bed load; and then, X_m is not necessarily identical to X_p value because the phenomenon is almost independent of X value. On the other hand, if X_p is smaller than 70, the suspended sediment cannot be neglected and the phenomenon is closely related to X value. Hence X_m should be chosen nearly to X_p as much as possible. Under the aforementioned condition, we can choose X_r value. Then, the scale of bed materials can be related to the vertical flow scale as follows:

$$\frac{U_r d_r}{\left[\left(\frac{C}{\sqrt{g}}\right)_r \nu_r\right]} = X_r \tag{5}$$

$$d_r = X_r^{6/7} h_r^{-2/7} \tag{6}$$

in which the Manning-Strickler equation has also been applied conveniently. The "distortion ratio", ϵ_d, defined as (d_r/h_r), is, therefore, determined this equation, and then the bed material to be used in the model should be chosen so that its relative density satisfies the following equation:

$$\left(\frac{\sigma}{\rho-1}\right)_r = \epsilon_d^{-2/3} \tag{7}$$

Practically, the laboratory size determines the horizontal scale x_r, and thus everything should be considered under such a constraint. Considering the condition that the flow in a model is also turbulent, the vertical flow scale, h_r, is determined and then, considerating equations (5) and (6), the distortion ratio and the relative density of model bed material should be determined simultaneously.

Next, we have to take care that the time scale of flow dynamics and that of fluvial process are generally quite different from each other. The former is simply determined based on the Froude number similarity law as follows:

$$t_r = x_r h_r^{-1/2} \tag{8}$$

On the other hand, the latter should be determined by using the governing equation of fluvial process. When the X-Y values similarity is established, the similarity as for sediment transport rate is kept and it means that $\Phi_B = 1$, in which Φ_B = dimensionless sediment transport rate and it is a unique and universal function of X and Y. According to the definition of Einstein's "transport intensity" (1942):

$$\Phi_B = \frac{qB}{\sqrt{\frac{\sigma}{\rho-1} g d^3}} \tag{9}$$

in which qB = sediment transport rate (substantial volume per unit width). Therefore, we obtain the following scale for sediment transport rate after substituting equation (7) into the relation that $\Phi_B = 1$.

$$qB_r = h_r^{1/3} d_r^{7/6} = \epsilon_d^{7/6} h_r^{3/2} \tag{10}$$

This gives the relation between the scale of sediment transport rate and the geometric scales.

The governing equation of fluvial process is the continuity equation of sediment (see equation [11]), and thus we obtain the time scale of fluvial process which is represented by T_r in order to distinguish from the time scale of flow dynamics t_r, as follows:

$$\frac{\partial z}{\partial t + \left[\frac{1}{1-\lambda}\right]} \frac{\partial qB}{\partial x} = 0 \tag{11}$$

$$T_r = \frac{(1-\lambda)_r x_r h_r}{qB_r} \tag{12}$$

$$T_r = x_r h_r^{-1/2} (\epsilon_d)^{-7/6} = x_r h_r^{3/2} d_r^{-7/6} \tag{13}$$

in which λ = porosity of sand and it is not so affected by scaling. As a conclusion as indicated by equation (13), the time scale of flow dynamics and that of fluvial process are quite different from each other when a distorted model is applied, and the ratio of these two scales (T_r/t_r) is $(\epsilon_d)^{-7/6}$.

An example is presented below: In the prototype, $U_p = 2.0$m/s; $h_p = 1.0$m; B_p = 40m (B = channel width), $d_p = 1.0$cm; and $(\sigma/\rho-1)_p = 1.65$ (the subscript p indicates the values corresponding to the prototype and the subscript m will represent the model values). We want to conduct a distorted model test where $x_r = 1/40$ and $h_r = 1/20$.

From the Froude number similarity, we obtain: $U_m = 44.7$ cm/s; $h_m = 5.0$ cm; $B_m = 1.0$ m and $Q_m = 22.4$ lit/s (Q = flow discharge). Since $X_p = 1200 >> 70$, it should be that $x_m > 70$. Here, we select $\epsilon_d = 5.0$ ($d_r = 1/4$). From equation (7), $(\sigma/\rho-1)_r$ = 1/2.92, then $(\sigma/\rho-1)_m = 0.56$. We can use "anthracite" as model bed materials, and d_m = 0.25 cm. Then, equation (10) gives us that $qB_r = 1/23.7$; and equation (13) gives that $T_r = 1/58.5$, while $t_r = 1/8.94$. Pay attention to the fact that the time scales for flow dynamics and fluvial process are actually quite different from each other.

3. SIMILARITY LAW FOR LOCAL SCOUR

The final or equilibrium scour depth around a circular cylinder is roughly related to the governing parameters as written below:

$$\zeta_f = \frac{z_{sf}}{D} = \frac{\sqrt{\gamma_0 \eta} - 1}{k_\omega} \quad \text{(for c.w.s.)} \tag{14}$$

$$\zeta_e = \frac{z_{se}}{D} = 1.2 \simeq 1.5 \quad \text{(for s.w.c.s.m.)} \tag{15}$$

in which z_s = scour depth; D = pier diameter; $\eta = \tau^*/\tau^*_c$; τ^* = Y = dimensionless bed shear stress; τ^*_c = dimensionless critical bed shear stress; k_ω = 1/7; γ_0 = 1.5; and "c.w.s." and "s.w.c.s.m." mean "clear water scour" and "scour with continuous sediment motion." As shown in Figure 2 originally prepared by Carstens (1966), it is obvious that there are two categories of scouring: clear water scour and scour with continuous sediment motion. Particularly in the case of clear water scour, the final scour depth obviously increases with the approaching flow intensity, and for example, equation (14) has been proposed (Tsujimoto-Nakagawa, 1986). The characteristic parameter η is a function of relative flow depth Z as shown below:

$$\eta = \frac{\tau^*}{\tau^*_c} = \left\{ \frac{Fr^2}{\left[\tau^*_c (7.66)^2 \left(\frac{\sigma}{\rho - 1} \right) \right]} \right\} Z^{-2/3} \tag{16}$$

in which Fr = Froude number; and 7.66 is a constant involved in the Manning-Strickler equation. Obviously, Z is related to the distorted ratio ϵ_d.

By the way, $\epsilon_d = 1$ is a kind of definition of "distorted model" used here. If η_r = 1 provides $(z_s/D)_r$ = 1 by some adjustment of other parameters, this distorted model is quite "exactly designed." In this case, the model bed materials to satisfy the following equation should be used.

$$\left(\frac{\sigma}{\rho - 1} \right)_r = (\tau^*_c)_r^{-1} \epsilon_d^{-2/3} \simeq \epsilon_d^{-2/3} \tag{17}$$

Then, the scale of scour depth is quite similar to the scale of pier diameter, and we can easily predict the actual scour depth in the prototype through a model test.

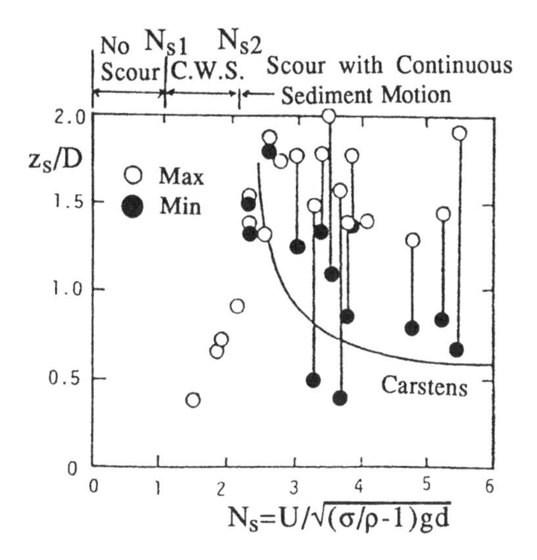

Figure 2. Clear water scour and scour with continuous sediment motion.

However, we often use the following not-true or wrong distorted model for the sake of convenience in the laboratory, because in general it is difficult to prepare a large amount of bed materials (light materials) except for sands though we cannot apply an undistorted model in order to keep the similarity on the pattern of sediment motion. However, even if one uses this kind of "not exactly distorted model," we can devise to evaluate the true value of scour depth in the prototype, as follows: When $\epsilon_d \neq 1$ but $(\sigma/\rho-1)_r = 1$, it concludes that $(z_s/D)_r \neq 1$. However, $(z_s/D)_r$ can be expressed by the following equation if we have the governing equation for scour depth based on a reasonable description of its mechanism such as equation (14).

$$\left(\frac{z_s}{D}\right)_r = \left[\frac{k}{k-1}\right]\epsilon_d^{-1/3} - \left[\frac{1}{k-1}\right] \tag{18}$$

in which

$$k = \left[\frac{\sqrt{\dfrac{\gamma 0}{\tau^*_c}}}{7.66\sqrt{\dfrac{\sigma}{\rho-1}}}\right]\cdot \mathrm{Fr}\cdot\left(\frac{h}{d}\right) \tag{19}$$

This gives us the relation between $(z_s/D)_r$ and ϵ_d. Figure 3 shows the experimental data as for the relation between $(z_s/D)_r$ and ϵ_d in "distorted model test without adjustment of the relative density of bed materials," collected by Nakagawa-Suzuki (1975). The theoretical curves and the experimental data show fairly good agreements.

	U_0 (cm/s)	D=20cm	D=8cm
○	undistorted	d=0.052cm	d=0.07cm
■	29.71		0.027cm 0.100cm
▼	23.32		0.027cm 0.100cm
□	14.43		0.027cm
▲	28.28	0.027cm 0.063cm 0.100cm	

Figure 3. Relation between $(z_s/D)_r$ and ϵ_d.

4. SUBSIDIARY EFFECTS ON LOCAL SCOUR

4.1. Analytical Modelling of Scouring Process at Front Foot of a Pier

Previously, several analytical models were derived to describe scouring process and to predict the scour depth. Laursen (1963), Carstens (1966), and others made analytical models using somehow macroscopic control volume as shown in Figure 4, but in this case it was difficult to use a reasonable sediment transport formula because of the spatial variations of flow and sediment behaviours in the control volume. On the other hand, Shen-Schneider-Karaki (1969) suggested how to evaluate the vortex as a main cause of local scour around a pier using the concept of the conservation law of the circulation. Based on their idea, the bed shear stress due to the vortex was evaluated, and an analytical model to describe scouring process by using the relation between bed shear stress and sediment pick-up rate established by Nakagawa-Tsujimoto (1980) was proposed (Tsujimoto-Nakagawa, 1986), in which a more microscopic control volume was used to consider the continuity of sediment and sediment behaviour as shown in Figure 5. Although this model focussed on only the stagnant plane as shown in Figure 5, the applicability of this model was supported by the facts that the maximum scour depth appears at the front foot of the pier and that it governs the whole scour hole development because of the geometrical similarity of the scour hole.

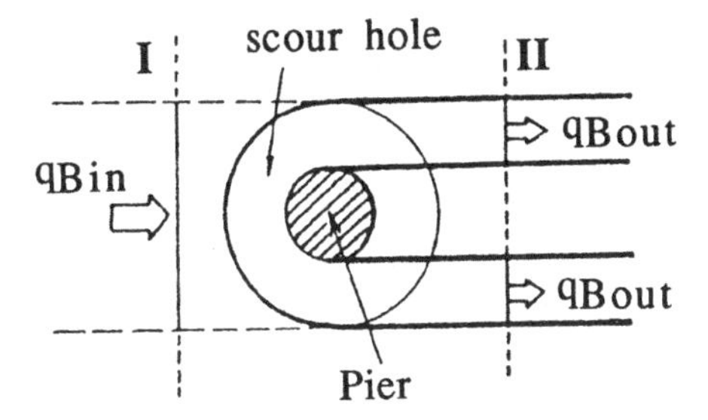

Figure 4. Macroscopic control volume for analytical model of local scour.

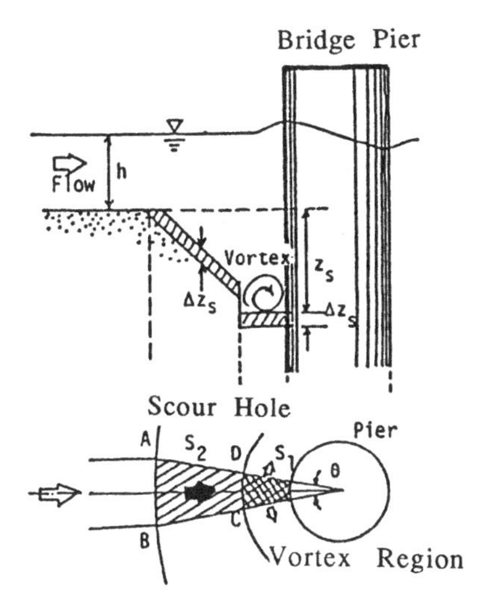

Figure 5. Microscopic control volume for analytical model of local scour.

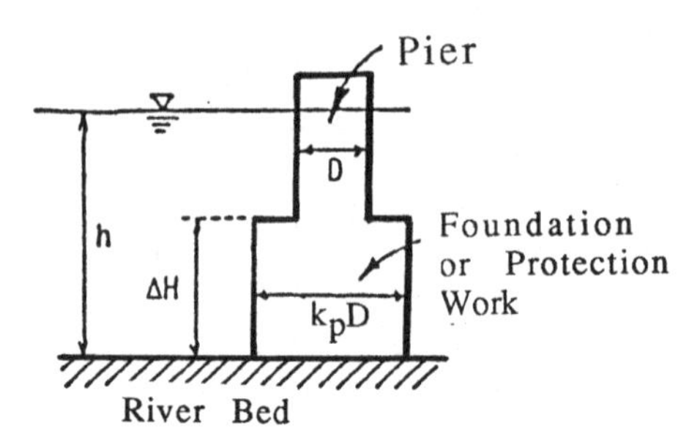

Figure 6. Modelling of supplementary structure as double-scale pier.

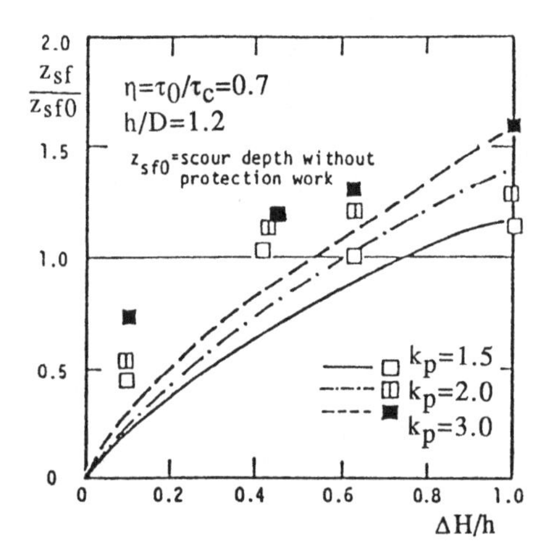

Figure 7. Effect of double-scale of pier on scour depth.

4.2. Effect of Supplementary Structure

The effect of supplementary structures of a bridge pier on scouring process is here discussed. From Photos 1-4, some of the bridges with supplementary structures are idealized as a pier with double-scale as shown in Figure 6. By the way, we have a mathematical model to represent the essential mechanism of the local scour (Tsujimoto-Nakagawa, 1986). We can easily modify this analytical model for single-scale pier to double-scale pier, and we are able to evaluate the effect of the supplementary structure as shown in Figure 7 (Tsujimoto, et al., 1987). In other words, such a modification for application is the most preferable requisite condition for an analytical or mathematical model. In Figure 7, the variation of the final scour depth around a double-scale pier from that of a single pier is shown against the scales of the supplementary sizes, and the predicted variation based on the analytical model shows a good agreement with the experimental results.

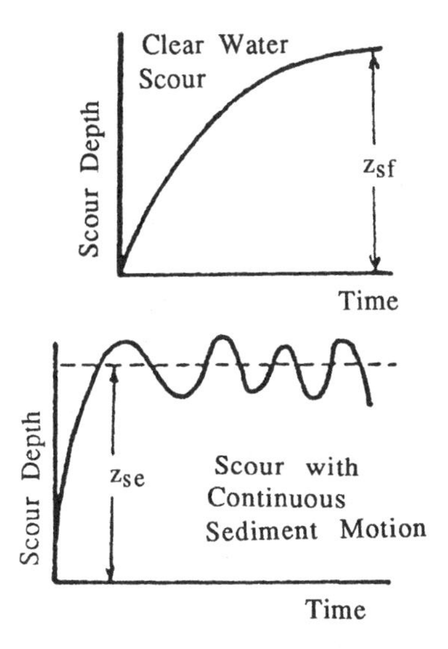

Figure 8. Clear water scour and scour with continuous sediment motion.

4.3. Effect of Dune Migration

Similarly, if we have a simple but essentially reasonable analytical model to describe scouring process around a pier, we are also able to describe the fluctuation of scour depth (see Figures 2 and 8) due to the migration of dunes (Tsujimoto-Nakagawa, 1986). Previously, most of researchers and engineers regarded the maximum scour depth as a sum of the ordinary scour depth (predicted without considering the effect of dunes) and dune height or somehow down-filtered dune height (Shen et al., 1969). However, now that we have an analytical model to describe scouring process, we can more physically reasonably predict the maximum scour depth also based on more statistically reasonable sense (see Figure 9, where the standard deviation of the fluctuating scour depth σ_Z is related to the standard deviation of duned bed elevation σ_Y; and the experimental data of this figure were collected by Suzuki et al. (1983)). Furthermore, according to this approach, we can predict the spectral properties of scour depth fluctuation if we know the sand wave spectra. In Figure 10, $S_Z(f)$ = frequency spectrum of the fluctuating scour depth; $S_Y(f)$ = frequency spectrum of sand waves; and f = frequency. When $S_Y(f)$ is proportional to f^{-2} as clarified by Hino (1968) and others, $S_Z(f)$ becomes proportional to f^{-4}, and it can be explained based on an analytical model (Tsujimoto-Nakagawa, 1986). This statistical information as for the fluctuation of scour depth may become more important for maintenance of a pier and designing protection works around a pier.

4.4. Effect of Sediment Sorting for Graded Bed Materials

In a model test of graded bed materials, the finer parts of the materials cannot be reproduced with keeping the similarity of the pattern of sediment motion between prototype and model. Hence, we often use a model test using a uniform sand and then we have to evaluate the effect of gradation of bed materials anyway.

According to the laboratory experiments, after scour hole is formed, the sediment sorting is remarkably observed. In region A of Figure 11 (the experiments were conducted by Nakagawa-Suzuki, 1977), at the front foot of the pier, where the maximum scour depth is finally achieved, the so-called armouring is recognized. And the variation of bed materials compositions at other typical points were also investigated experimentally, and the outline of the results are illustrated in Figure 11. Rigorously, we can see the temporal variation of gradation curves of bed materials in this figure. From the figure we can understand that the local scour in case of sediment mixtures is inevitably accompanied with local sediment sorting.

Since we have an analytical model to describe scouring process and one to describe sediment motion for each grain size of sand mixtures and the subsequent temporal variation of bed surface constitution (Nakagawa-Tsujimoto-Hara, 1977), we can evaluate the difference of the progress of scouring and the final scour depth in case of graded bed materials and those in case of uniform sand, and we are able to predict the prototype scour behaviour from the model test using uniform material. Such a technique is quite possible, and Figure 12 is an example of comparison of the analytical prediction with the experimental data. The data were obtained by Nakagawa-Otsubo-Nakagawa (1981) using a rectangular pier. In this figure, we can recognize that the progress of scouring is more retarded for bed materials with larger value of $\sigma_d = \sqrt{d_{84}/d_{16}}$, and that the final scour depth decreases against it. Furthermore, these characteristics can be evaluated or described by a derived analytical model as represented by some calculated curves compared with the experimental data in Figure 12.

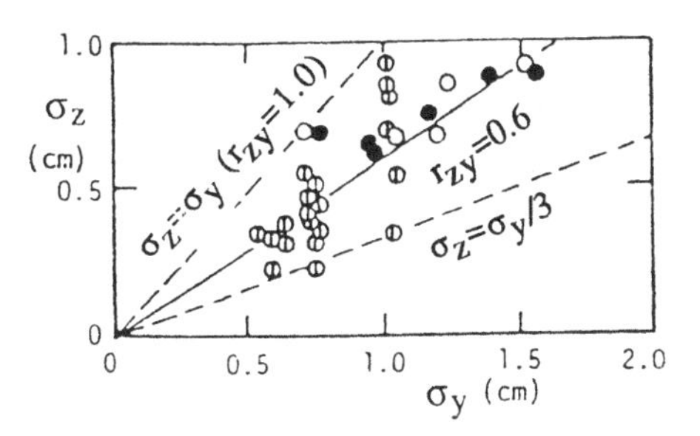

Figure 9. Standard deviation of fluctuating scour depth due to dune migration.

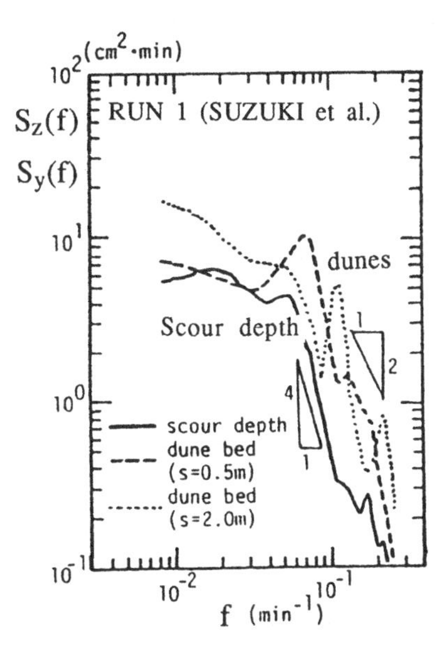

Figure 10. Spectra of sand waves and fluctuating scour depth.

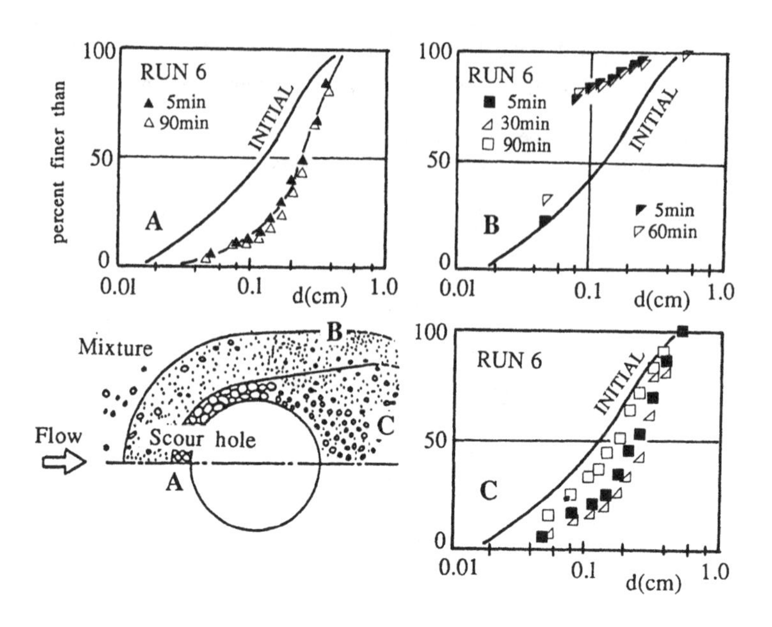

Figure 11. Sediment sorting accompanying local scour around a pier.

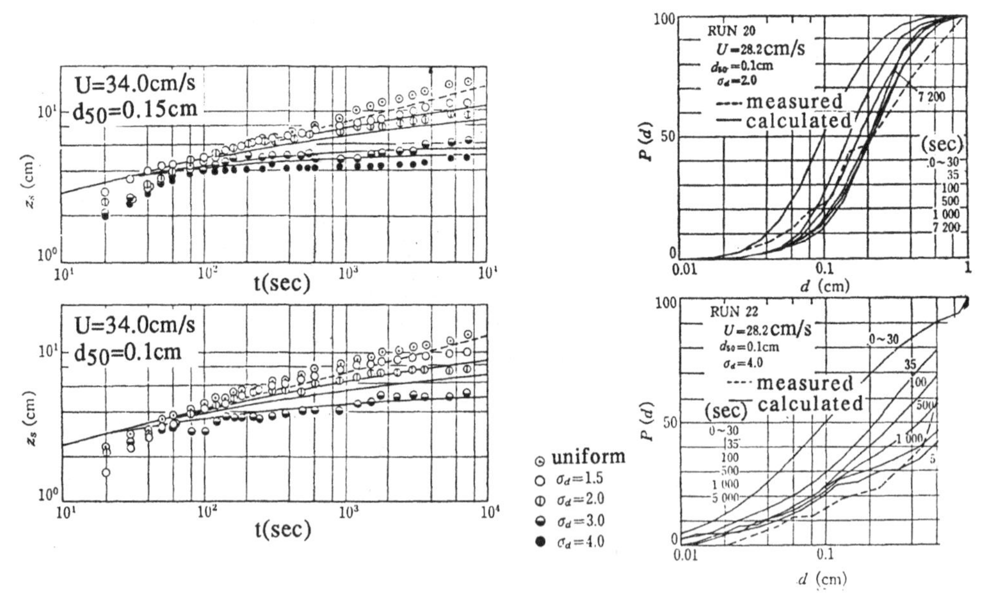

Figure 12. Scour depth reduction due to gradation of bed materials.

Photo 1. Sakura Bridge (Kino River).

Photo 2. Omonkuchi Bridge (Kino River).

Photo 3. Ryumon Bridge (Kino River).

Photo 4. J.N.R. Iwade Bridge (Kino River).

5. CONCLUSIONS

In general, fluvial processes involve several scales, which are quite different from each other in size, and thus, some scales inevitably become too small in a small model. In such a case, we have a technique of "distorted model" tests, in which not only the vertical scale but also the scale of sand diameter have different scales in some cases. In this paper, the theoretical backgrounds are explained with particular reference to local scour around a bridge pier, and it is emphasized here that the time scale of fluvial process is quite different from that of fluid motion, particularly in the case of a distorted model.

However skillfully one operates a distorted model test, still some subsidiary effects cannot be simply reproduced in a small and/or distorted model. For example, the effects of the supplementary structure, dune migration, and the gradation of bed materials on local scour around a bridge pier have been discussed. These effects cannot be easily reproduced in a physical model, but if we have an analytical model of the fundamental phenomena based on their essential mechanisms, they are very helpful to evaluate these subsidiary effects. In this paper, some examples are introduced to illustrate the importance of skillful manipulation of these two approaches: distorted physical modelling and analytical modelling to support the former. This kind of sophisticated approach may provide a better prediction of the actual phenomena in prototype than only using a physical model test where almost all of subprocesses somehow complicatedly related to the phenomena are most carefully or even completely tried to be reproduced.

6. REFERENCES

Carstens, M.R. (1966). Similarity laws for localized scour. *J. Hydraulics Division*, ASCE, Vol 92, HY3, 13-36.

Hino, M. (1968). Equilibrium range spectra of sand waves. *J. Fluid Mechanics*, Vol 34, Part 3, 565-573.

Laursen, E.M. (1963). An analysis of relief bridge scour. *J. Hydraulics Division*, ASCE, Vol 89, HY3, Proc. Paper 3516, 93-118.

Nakagawa, H., Otsubo, K., and Nakagawa, M. (1981). Characteristics of local scour around bridge piers for nonuniform sediment. *Proc. JSCE*, No 314, 53-65 (in Japanese).

Nakagawa, H., and Suzuki, K. (1975). Characteristics of local scour around bridge piers by tidal currents. *Proc. 22nd Japanese Conference on Coastal Engineering*, JSCE, 21-27 (in Japanese).

Nakagawa, H., and Suzuki, K. (1975). Armoring effect on local scour around bridge piers. *Annuals, Disaster Prevention Research Institute*, Kyoto University, No 18B, 689-700 (in Japanese).

Nakagawa, H., and Tsujimoto, T. (1980). Sand bed instability due to bed load motion. *J. Hydraulics Division*, ASCE, Vol 106, HY12, 2029-2051.

Nakagawa, H., Tsujimoto, T., and Hara, T. (1977). Armoring of alluvial bed composed of sediment mixtures. *Annuals, Disaster Prevention Research Institute*, Kyoto University, No 20B-2, 355-370 (in Japanese).

Shen, H.W. (1986) Scour near piers. *River Sedimentation* (H. W. Shen, ed.), Vol 2, Chapter 23, Water Resources Publications.

Shen, H.W., Schneider, V.R., and Karaki, S. (1969). Local scour around bridge piers. *J. Hydraulics Division*, ASCE, Vol 95, HY6, 1919-1940.

Suzuki, K., Michiue, M., and Kataoka, K. (1983). Influence of sand waves on the local scour around a bridge pier. *Proc. 27th Japanese Conference on Hydraulics*, JSCE, 659-664 (in Japanese).

Tsujimoto, T., Murakami, S., Fukushima, T., and Shibata, R. (1987). Local scour around bridge piers in rivers and its protection works. *Memoirs, Faculty of Technology, Kanazawa University*, Vol 20, No 1, 11-21.

Tsujimoto, T., and Nakagawa, H. (1986). Physical modelling of local scour around a bridge pier and prediction of fluctuation of scour depth due to dune migration. *Proc. IAHR Symposium on Scale Effects in Modelling Sediment Transport Phenomena*, IAHR, Toronto, Ontario, Canada.

SCALE EFFECTS IN THE REPRODUCTION OF THE OVERALL BED TOPOGRAPHY IN RIVER MODELS

N. STRUIKSMA
Senior Project Engineer
Delft Hydraulics
P.O. Box 152
8300 AD Emmeloord
The Netherlands

ABSTRACT. In river scale models with movable bed a significant scale effect occurs due to the distortion of the model. This distortion is necessary in order to reproduce the flow pattern in a similar way (roughness condition). An attempt is made to clarify and quantify this scale effect by using experiences obtained during the development of a mathematical model for the computation of the 2-D bed topography. From a linear analysis of the steady state of this model an important parameter governing the interaction between the water and sediment movement could be defined which cannot be reproduced in a correct way. The parameter is used to quantify the scale effect in terms of wave length and damping of the 2-D bed deformation.

1. INTRODUCTION

Most scaling procedures for river models with movable bed aiming at a simulation of the bed topography are based on an analysis of 1-D equations of water and sediment motion. In spite of this simple approach, conflicts are generally present by applying the resulting scale relations. If these procedures are extended to an analysis of the 2-D equations in the horizontal plane, then some additional conflicts arise which decrease the reliability of scale model investigations more. This 2-D analysis is facilitated due to the significant progress in the last decade of 2-D mathematical modelling for the computation of the flow field and bed topography (Engelund, 1974 and Struiksma et al; 1985). Using the experience obtained from this development interesting conclusions can be drawn about scaling conflicts which is the topic of this paper.

For simplicity, the subject matter is restricted to the scaling of meandering rivers with relatively stable banks and shallow flow. Additionally, only the equilibrium bed topography is considered and it is assumed that:

H. W. Shen (ed.), Movable Bed Physical Models, 49–58.

- the Froude Number is small to moderate (also in the model),
- the horizontal exchange of momentum in the vertical planes due to shear and spiral flow does not affect the flow pattern,
- the spatial variation of the hydraulic roughness (Chézy coefficient) can be neglected,
- the rate of sediment transport can be determined by local conditions (dominant bed load), and
- grain sorting is insignificant (uniform bed material).

Hereafter the scale n_x of any parameter X is defined as the ratio of prototype to model value ($n_x = X_p/X_m$). The physical phenomena are described in an orthogonal curvilinear coordinate system based on the depth-averaged flow field as indicated in Figure 1.

2. WATER MOVEMENT

As the bed topography depends on the flow pattern and the other way around it is necessary to aim at a similar reproduction of this pattern.

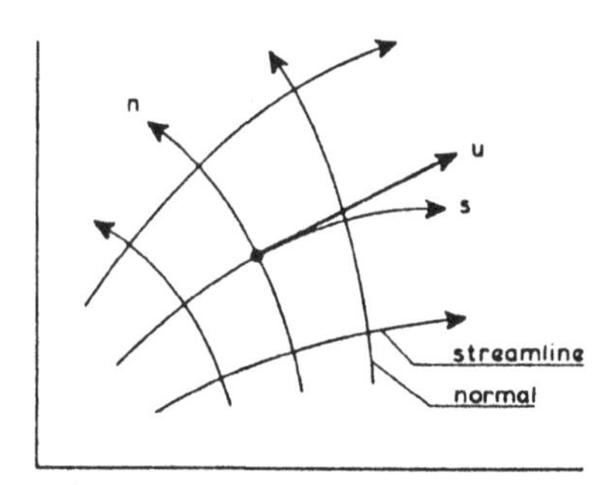

Figure 1. Coordinate system.

The flow as indicated can be characterized by the Reynolds number Re = uh/ν, Froude Number $F = u/\sqrt{gh}$ and the parameter gL/C^2h, in which u is the depth-averaged flow velocity which defines the main flow direction and s-coordinate, h is the water depth, ν is the kinematic viscosity, g is the acceleration due to gravity, C is the hydraulic bed roughness coefficient (Chézy) and L is a characteristic length (for instance, the meander length).

To provide a sufficient reproduction of turbulence in a model, the Reynolds Number has to exceed a certain value. Generally, in a scale model with a movable bed this requirement can be fulfilled easily. The importance of the Froude Number and parameter gL/C^2h appears from an analysis of, for instance, the dimensionless vorticity equation describing the steady depth-averaged flow field in the coordinate system of Figure 1, which reads (De Vriend and Struiksma, 1983):

$$\frac{\partial w'}{\partial s'} + \omega' \left(2 \frac{gL}{C^2H} - \frac{1}{h'} \frac{\partial h'}{\partial s'} \right) = \frac{gL}{C^2H} \frac{1}{h'} \left(\frac{u'}{R'} + \frac{u'}{h'} \frac{\partial h'}{\partial n'} \right) \tag{1}$$

in which $w' = \partial u'/\partial n' + u'/R'$ is the dimensionless vorticity, $s' = s/L$ and $n' = n/L$ are the dimensionless coordinate, $u' = u/U$ is the dimensionless flow velocity, $h' = h/H$ is the dimensionless water depth, $R' = R/L$ is the dimensionless radius of curvature of the stream lines (positive as the normal lines diverge), and L, H and U are characteristic measures for length, depth and velocity, respectively.

Order of magnitude estimates of the factors in damping and production terms of Equation 1 show that the Froude Number is of secondary importance for the flow as indicated before. It governs the deformation of the water surface which is small compared with that of the river bed. This gives the freedom to apply $n_F < 1$ provided that this leads to not too large Froude Numbers in the model. The freedom eases the selection of the bed material in a model. The consequences of the too steep water and bed surface gradients generated in this way are compensated by a tilting of the datum of the model in main flow direction.

The dominating factors in Equation 1 determining the behavior of the vorticity are the bed topography represented by the water depth and the parameter gL/C^2H. Reproducing the flow pattern in a similar way implies that n_H has to be invariable in space and the parameter gL/C^2H is on scale 1. The latter condition gives the so-called roughness condition (Jansen, 1979):

$$n_C^2 = \frac{n_L}{n_H} \tag{2}$$

For alluvial roughness this condition will lead in most cases to distorted models due to relatively large bed roughness in the model ($n_C > 1$).

As has been shown by for instance De Vriend (1981) and Rozovskii (1961) the influence of spiral motion (generated by curved streamlines) on the main flow pattern can be remarkable. It results in an extra concentration of flow along the bank of an outer bend. This effect is not incorporated in Equation 1. Here suffice it to say that the effect is exaggerated in distorted models (scale effect).

3. SEDIMENT MOTION AND BED TOPOGRAPHY

A similar reproduction of the bed topography can only be achieved if the sediment transport scale is invariable in space. In fact it implies that the sediment motion pattern along the bed is reproduced similarly. If the sediment motion direction is characterized by the angle α with the s-coordinate (counter clockwise is positive) the condition is identical to:

$$n_{\tan\alpha} = 1 \tag{3}$$

in which $\tan\alpha = s_n/s_s$, s_n and s_s are the sediment transport including pores per unit length in n- and s-direction, respectively.

If it is assumed that the sediment transport s_s is varying according to:

$$s_s \propto u^b \tag{4}$$

then it can be expected that the condition might be fulfilled if the value of the exponent b in the model is equal to that in the prototype (the exponent b may vary in space). For this it is assumed further that a unique function (sediment transport formula) exists between the transport parameter $\Psi = s_s/\sqrt{D^3 \Delta g}$ and the Shields parameter $\theta = u^2/C^2 \Delta D$, in which D is the grain size and Δ is the relative submerged density of the sediment. Then to arrive at $b_m = b_p$ it is stated that :

$$n_\theta = 1 \quad \text{or} \quad n_u = n_C \sqrt{n_{\Delta D}} \tag{5}$$

which will be called hereafter the "ideal" velocity scale (Jansen, 1979).

The question then arises if it is possible to fulfill the ideal velocity scale condition. For an answer the particle path direction tan α will be considered in more detail. A composition of forces acting on particles moving along a mildly sloping bed leads approximately to the following dimensionless equation (Struiksma et al., 1985):

$$\tan \alpha = - A \frac{H}{L} \frac{h'}{R'} + \frac{H}{f(\theta) L} \frac{\partial h'}{\partial n'} \tag{6}$$

in which A is a coefficient resulting from the depth integration of the flow. The coefficient depends slightly on the hydraulic bed roughness which will be ignored. The coefficient weighs the influence of the spiral motion whereas the f(θ) weighs the influence of the sloping bed in normal (transverse) direction on the transport direction. For a review on f(θ) reference is made to Odgaard (1981). If it is assumed that the function f(θ) is unique than from Equation 6 with $n_A = 1$ and $n_\theta = 1$, Equation 5, it is found that:

$$n_{\tan \alpha} = n_H / n_L \tag{7}$$

For distorted models this is in conflict with the ideal velocity scale condition ($n_{\tan \alpha} = 1$). Consequently this leads to scale effects at places where tan α is relatively large. The conflict has also an implication for the morphological time scale. In fact this scale cannot be defined anymore because a distinction has to be made between the transverse (normal) and main-flow direction. The conflict cannot be avoided because the roughness condition, Equation 2, has to be fulfilled (Struiksma, 1980).

For the special case of fully developed flow in a "long" circular bend with constant width tan α becomes negligible (particle path parallel to the banks). Then Equation 6 transforms into:

$$\frac{\partial h'}{\partial n'} = A \; f(\theta) \; \frac{h'}{R'} \tag{8}$$

It can be seen easily that the assumption made about the function f(θ) together with $n_A = 1$ and $n_\theta = 1$ leads to a similar reproduction of the fully developed transverse bed slope.

4. INTERACTION OF WATER AND SEDIMENT MOTION

Due to the significant progress made in 2-D mathematical modelling of the bed topography, it is possible now to estimate roughly the consequences of the scale conflict described in the foregoing chapter. Use will be made of the findings at the Delft Hydraulics Laboratory during the last decade. These findings can be found in Struiksma (1985), De Vriend and Struiksma (1983) and Struiksma et al., (1985). The efforts made resulted ultimately in a mathematical model which incorporates the following physical phenomena:

- steady water motion,
- depth-averaged main velocity, including inertia, bottom shear stress and (crudely approximated) secondary flow convection,
- logarithmic vertical distribution of the main flow,
- spiral flow intensity, including inertial grow and decay,
- vertical distribution of spiral flow as in fully developed curved flow with a logarithmic main velocity profile,
- magnitude and direction of bed shear stress according to fully developed curved flow,
- rate of the sediment transport with a formula expressed in terms of the total bed shear stress and, if necessary, corrected for slope effects,
- direction of sediment transport influenced by direction of bed shear stress and sloping bed, and
- time-dependent bed level based on local sediment balance.

From a linear analysis of the steady state of the model (Struiksma et al., 1985) it appeared that the main features of the bed deformation can be described by the zero-order solution according to the fully developed bed deformation, Equation 8, which is determined by local parameters only and a wave-like first-order solution around the zero-order solution. The latter is caused by the retarded reaction of the water and sediment motion to changes of the zero-order solution. It can be characterized by a wave length and damping length which are governed by the interaction parameter:

$$\frac{\lambda_s}{\lambda_w} = \frac{2}{\pi^2}\,\frac{g}{C^2}\left(\frac{B}{h}\right)^2 f(\theta) \tag{9}$$

in which $\lambda_s = \pi^{-2}\, h\, f(\theta)\, (B/h)^2$ is the relaxation length of the bed deformation, $\lambda_w = C^2 h/2g$ is the relaxation length of the main flow, and B is the width of the river.

Another important factor is the exponent b in the sediment transport formula, Equation 4. In Figure 2 for b = 5 and for λ_s/λ_w ranging from 0.2 to 5 the relation between dimensionless wave numbers and interaction parameter is shown according to:

$$\frac{2\pi}{L_p}\,\lambda_w = \frac{1}{2}\sqrt{(b+1)\,\frac{\lambda_w}{\lambda_s} - \left(\frac{\lambda_w}{\lambda_s}\right)^2\left(\frac{b-3}{2}\right)^2} \tag{10}$$

and

$$\frac{\lambda_w}{L_D} \approx \frac{1}{2}\left(\frac{\lambda_w}{\lambda_s} - \frac{b-3}{2}\right) \tag{11}$$

in which L_p is the wave length (meander length approximately) and L_D is the damping length.

From these equations the significant influence of the exponents b can be found easily.

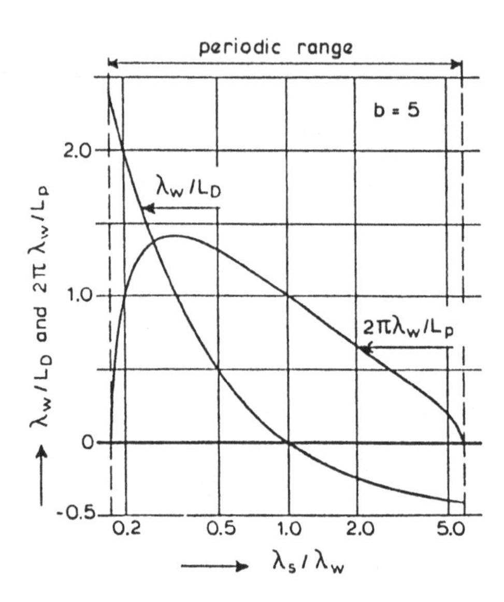

Figure 2. Wave length and damping length (Struiksma et al., 1985).

From Figure 2 it can be seen that for $\lambda_s/\lambda_w > 1$ the amplitude of the bed deformation will grow in downstream direction (negative damping length). Physically it implies that in the river, islands will be formed resulting in a braided river which will not be treated here.

The results of the linear analysis are confirmed by computations with the complete mathematical model which includes also non-linear effects. In Figure 3 an example for a circular curved flume is given of such a computation compared with measurements. Also the solution of Equation 8 is indicated. Clearly it can be observed that the result has the form of a damped wave around the axi-symmetrical solution according to Equation 8. This behavior is confirmed by the measurements. In the downstream straight section of the flume this behavior results in a "waggling" of the bed. Figure 3 confirms the relevance of the linear analysis embodied by Equations 10 and 11 and Figure 2. For that reason it will be used here to estimate roughly the scaling conflict as described in the foregoing chapter.

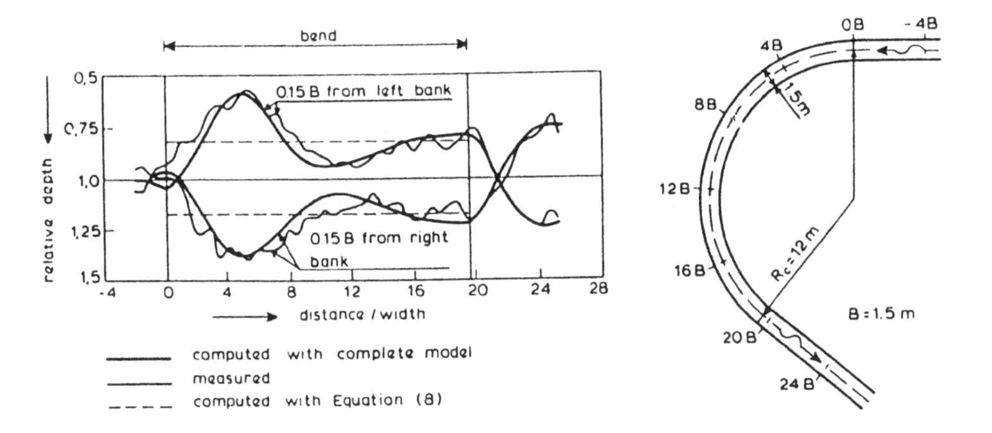

Figure 3. Measured and computed longitudinal bed profiles (Struiksma et al., 1985).

For a similar reproduction of the bed topography it is necessary that:

$$n_{L_p} = n_L \text{ and } n_{L_D} = n_L \qquad (12)$$

Together with $n_{\lambda_w} = n_L$ (roughness condition) this implies that the interaction parameter λ_s/λ_w, Equation 9, has to be reproduced on scale 1 or:

$$n_{\lambda_s} = n_L \qquad (13)$$

This is impossible without violating other important scale conditions. Generally $n_{\lambda_s} = n_L$ which results, according to Figure 2, in a shorter bed wave length and more damping (see Figure 4). In other words the bed topography in a scale model will be more dominated by Equation 8 than the bed topography in the prototype.

5. DISCUSSION

Maybe a partial escape is possible because there is some evidence from experiences with computation with the complete mathematical model that with respect to the bed slope weigh function $f(\theta)$ (and also the spiral flow weigh coefficient A) a distinction has to be made between prototype and laboratory conditions with a diminishing effect on the scaling conflict. However, lack of adequate prototype data prevents a definitive conclusion for the time being.

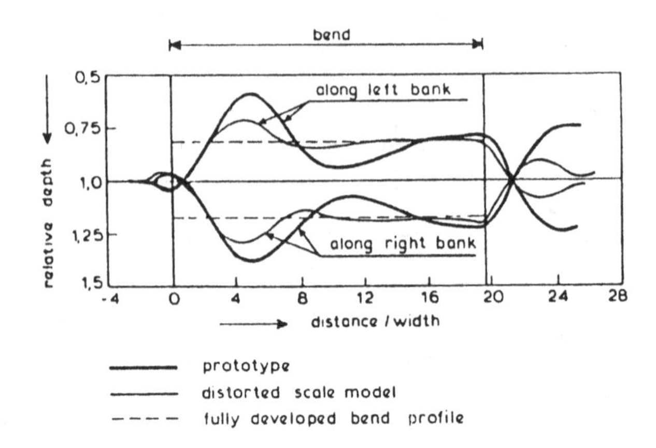

Figure 4. Scale effect in bend topography reproduction introduced by distortion of scale. model

Another possibility to escape is to distinguish between transverse and longitudinal length scales as was suggested by Engelund (1981). A disadvantage of this method is a significant mutilation of the planform of the prototype.

A delicate point of discussion is that during the calibration of the model it will generally be possible to arrive at a similar reproduction of the bed topography by adjusting the boundary conditions. For instance, in a distorted scale model the interaction parameter λ_s/λ_w is too small. Enlarging the Shields parameter θ by simply increasing the discharge without changing the depth gives due to the form of the weigh function $f(\theta)$ also an enlarging of the interaction parameter and the fully developed transverse bed slope according to equation 8. After some trial and error the object aimed at is gained. However, the scale effect described in the foregoing chapters is still present and this can lead during the progress of the investigation to wrong conclusions and recommendations.

Good prospects are present in modelling gravel bed rivers if the grain size can be scaled down on length scale. Then most probably the bed roughness scale will be close to 1 which leads via the roughness condition to a non-distorted model.

6. SUMMARY OF CONCLUSIONS

The conclusions concerning scale effects in models with movable bed can be summarized as follows:

- For the design of river scale models the Froude condition is of secondary importance. This gives the freedom to select $n_F < 1$ which is a great advantage because it facilitates the selection of other scales.
- A serious scale effect arises due to the fact that generally the alluvial bed roughness cannot be reproduced on scale 1 ($n_C > 1$). Then according to the roughness condition the model has to be distorted. In such a model it is not possible to reproduce the wave and damping length of the bed deformation on length scale.
- During the calibration of a model it is possible to arrive at a fair reproduction of the bed topography by adjusting the boundary conditions. However, the scale effects are obscured then because use is made of a groundless distortion of the ratio between the fully developed transverse bed slope and the overshoot phenomena embodied by the damped wave behavior of the bed deformation.

7. REFERENCES

Engelund, F. (1974), 'Flow and Bed Topography in Channel Bends,' *J. Hydr. Div., ASCE,* **100**, No. HY11, P. 1631.

Engelund, F. (1981), *Similarity Laws of Alluvial Streams*, Tech. Univ. Denmark, Prog. Rep. 55, 21.

De Vriend, H. J. (1981), 'Steady Flow in Shallow Channel Bends,' *Comm. on Hydr.* 81-3, Department of Civil Engineering, Delft Univ. of Techn., Delft, The Netherlands.

De Vriend, H. J. and Struiksma, N. (1983), *Flow and Bed Deformation in River Bends,* Presented at Rivers '83, ASCE, New Orleans, La.

Jansen, P. Ph. (1979), *Principles of River Engineering*, Pitman Publ. Ltd. London.

Odgaard, A. J. (1981),'Transverse Bed Slope in Alluvial Channel Bends,' *J. Hydr. Div.*, ASCE, **107**, No. HY12, 1677.

Rozoskii, I. L. (1961), *Flow of Water in Bends of Open Channels*, Israel Program for Scientific Translations, Jerusalem.

Struiksma, N. (1980), *Recent Developments in Design of River Scale Models with Mobile Bed*, IAHR Symp. on River Engrg. and its Interaction with Hydrological Research, Belgrado.

Struiksma, N. (1985), 'Prediction of 2-D Bed Topography in Rivers,' *J. of Hydr. Engrg.*, ASCE, **III**, No 8, 1169.

Struiksma, N., Olesen, K. W., Flokstra, C., and De Vriend, H. J. (1985), 'Bed Deformation in Curved Alluvial Channels,' *J. of Hydr. Res.*, IAHR, **23**, No.1.

List of Symbols

A	weigh coefficient of spiral flow intensity
B	width
b	exponent in simplified sediment transport formula
C	bed roughness coefficient (Chézy)
D	grain size
F	Froude Number
$f(\theta)$	weigh function of influence of bed slope
g	acceleration due to gravity
H	characteristic depth
h	wter depth
L	characteristic length
L_D	damping length of bed deformation
L_p	wave length of bed deformation
n_x	scale of parameter X
R	radius of curvature of streamline
Re	Reynolds Number
s,n	coordinates in main flow and normal direction, respectively
s_n, s_s	sediment transport including pores per unit length in n- and s-direction, respectively
U	characteristic velocity
u	depth-averaged main flow velocity
α	angle of sediment transport direction
Δ	relative submerged density of sediment
θ	Shields parameter
λ_s	relaxation length of bed deformation
λ_w	relaxation length of main flow
ν	kinematic viscosity
Ψ	sediment transport parameter
ω	vorticity
p,m	indices, denoting prototype and model, respectively

ON THE SCALING OF BRAIDED SAND-BED RIVERS

G. J. KLAASSEN
Delft Hydraulics
P. O. Box 152
8300 AD Emmeloord
The Netherlands

ABSTRACT. A scaling method is described that was developed recently to design a physical scale model of the Jamuna (= Lower Brahmaputra) River in Bangladesh. The model investigation is related to a proposed bridge across this braided river. The scaling method aims at a reproduction of the braiding pattern in order to study the effect of the proposed bridge on the channel characteristics. The scaling method itself, possible scale effects and some other aspects in designing and running such a model are discussed.

1. INTRODUCTION

More and more, mathematical models are taking over for problems where, in the past, scale models were usually applied. This holds also for certain types of movable bed problems, notably for rivers with fixed bank (see e.g. Struiksma, 1981). Certain movable bed river problems, however, cannot yet be tackled with mathematical models, notably problems related to channels where the banks are not fixed (regime channels) and channel patterns. In this paper a scaling procedure is discussed which was developed for a braided river, where reproduction of the channel pattern was the most important criterion. It is shown here that this results in a scaling procedure with a very limited degree of freedom; consequently considerable scale effects may be introduced.

The reference problem is a proposed bridge across the Jamuna River in Bangladesh (see Figure 1).

H. W. Shen (ed.), Movable Bed Physical Models, 59–71.

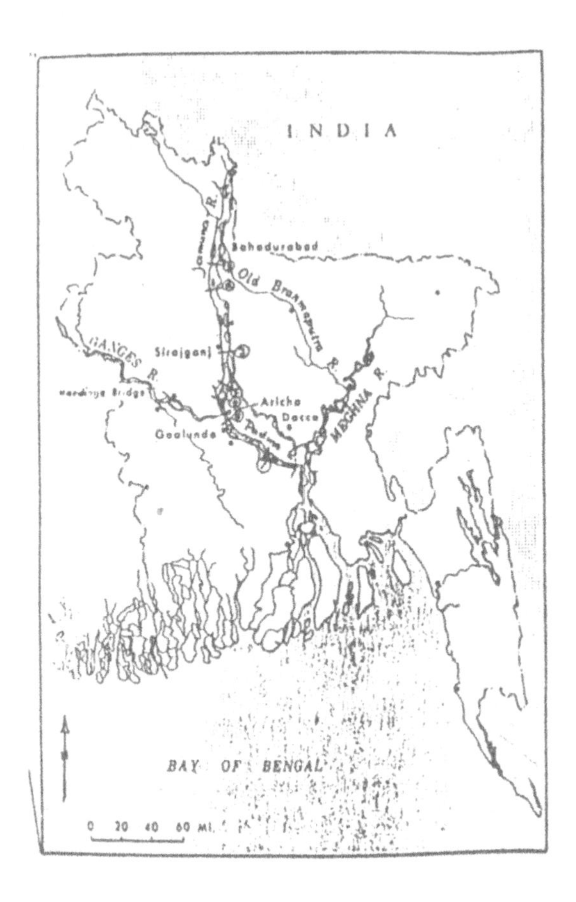

Figure 1. Map of Bangladesh with the Jamuna River.

The Jamuna River is the lowest reach of the Brahmaputra River. Figure 2 provides a 1:1,000,000 scale satellite image of the Jamuna River and of its confluence with the Ganges River. As can be observed, the Jamuna River is a braided river, with on the average three braids upstream of Sirajganj (near the proposed bridge site) and about two braids more downstream. The river bed can be characterized as a flat sand bed in which a number of deep channels are present. The shallow bars in between the channels are locally called "chars".

The main characteristics of the Jamuna Bridge are the following:

• total length in Bangladesh	~ 300 km
• maximum total width	~ 15 km
• number of main channels	2-3
• width of main channels during floods	~ 2.5 km
• average depth main channel (h)	~ 8 m
• average depth during low flow conditions	3 m
• average depth above chars during floods	1-2 m
• Chézy coefficient (average 1)	~ 70 $m^{\frac{1}{2}}/s$
• Chézy coefficient floods	~ 90 $m^{\frac{1}{2}}/s$
• D_{50}	~ 200 μm
• slope (i)	~ $0.7.10^{-4}$
• $\theta = hi/\Delta D_{50}$ (floods)	~ 1.7
• average velocity during floods	~ 2m/s
• maximum total discharge	~ 100,000 m^3/s

For more details on the river characteristics of the Jamuna River, reference is made to RPT/Nedeco/BCL (1987) or Klaassen et al. (1988).

It was proposed to do model investigations in a movable bed scale model. The purpose of this scale model investigation would be:

- to study the effect of the river training works on channel patterns;
- to identify "worst" channel patterns (for flood and low-flow conditions) for the river training works;
- to study the effect of gradual shifts of river upstream on channel patterns near bridge and training works.

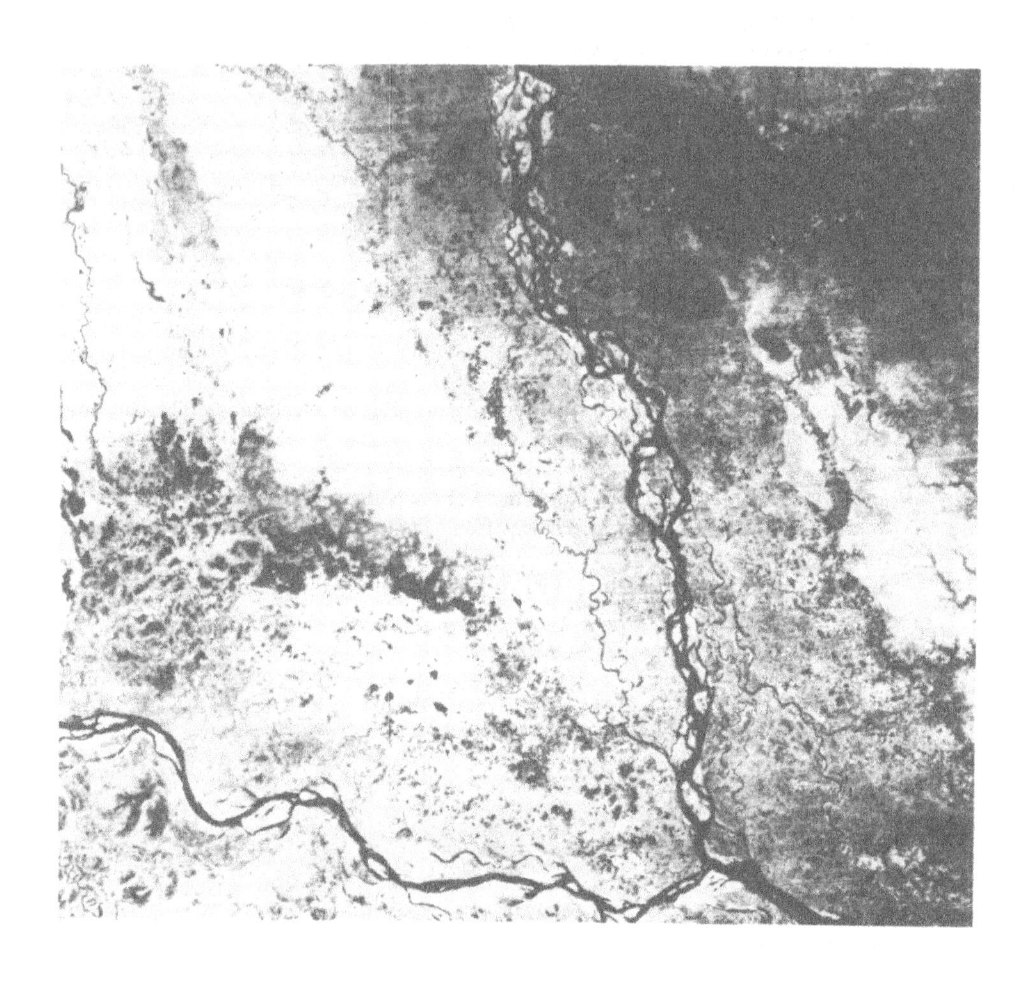

Figure 2. Satellite image of Jamuna River, February 2, 1973 (scale approximately 1:1,000,000)

Alternatively, historical channel patterns could be used to arrive at design conditions for the river training works required. From the results of an exploring model test, it was concluded, however, that the presence of a bridge constricting the river width has a significant effect on the channel patterns.

It should be stressed that the scale model discussed here is only one element of a number of research tools applied in this study. In addition the following investigations are foreseen:

- a 1-d mathematical model of the morphology of the Brahmaputra River to study long-term changes resulting from embankments, major diversion, dams, etc.;
- a 1-d mathematical model to study constriction scour, because the bridge opening will probably be only about 6 km;
- a 2-d mathematical model of the depth-averaged flow field underneath the bridge; this model will use the results of the movable-bed scale model as input and a curvilinear coordinate system will be used (Wijbenga, 1985);
- detail scale models of structures (guide bunds, groyns) that will use the results of the mathematical model of the flow field as boundary conditions;
- an extensive analysis of historical river data;
- an extensive study of satellite images to determine channel characteristics and celerities of channel pattern changes over the period 1973-1987;
- extensive hydrographic field measurements during the flood period in 1987.

2. SCALING PROCEDURE

2.1. Starting Points

The following parameters were considered to be important and less important, respectively, for the present scale model investigations:

Important:
- channel patterns (braiding with 2-3 channels)
- roughness condition (reproduction of eddies)
- width of channels reproduced on length scale
- ratio curvature/channel width

Less important:
- Re-condition ($Re_m > \Delta 10^3$)
- Fr-condition ($Fr_m < 1$)
- ideal velocity-scale
- mobility criterion (?)
- distortion
- horizontal celerity of channels

Other requirements:
- not too large (costs, celerity)
- high celerity of horizontal channel changes ("fast" model)
- sediment transport recirculated but measurable

2.2. Background of Scaling Procedure

The most important criterion which has to be used to derive scale rules for the present model is the reproduction of the channel patterns. Thus in the movable bed model anyhow a braided pattern should be present with two to three braids. No theoretical method is available in literature that allows to derive such a scale rule in a "scientific" way. However, there is some information in the literature that may be useful.

According to Leopold, Miller and Wolman (1964) a river is on the threshold between meandering and braiding (so on a transition from a single channel pattern to a multiple channel pattern) when:

$$i = \text{constant.}\ Q^{-0.44} \tag{1}$$

where i = stream gradient (-) and Q = annual flood (m^3/s). Channel patterns of braided streams have been studied by Howard et al. (1970). The find that the numbers of braids n is determined by:

$$n = 0.58\ R_q^{-0.19}\ i^{0.41} \cdot Q^{0.25} \tag{2}$$

where R_q = ratio of mean annual flood to the mean annual discharge. Equation 2 can also be written as:

$$i = \text{constant} \cdot R_q^{+0.46} \cdot Q^{-0.60} \tag{3}$$

an expression which has some similarity with Equation 1. Struiksma and Klaassen (1988) provide theoretical evidence for such a type of formula ruling channel patterns. They also indicate that from theoretical reasoning it follows that probably the ruling equation for the channel pattern reads as:

$$\frac{i}{\Delta D} = \alpha \cdot (Q)^{\beta} \tag{4}$$

where α and β still have to be defined and do not need to be constant over the full range of Q. This is in line with Ferguson (1984), who shows that i/D is preferable over i alone. Equation 4 is indicated in a schematic way in Figure 3:

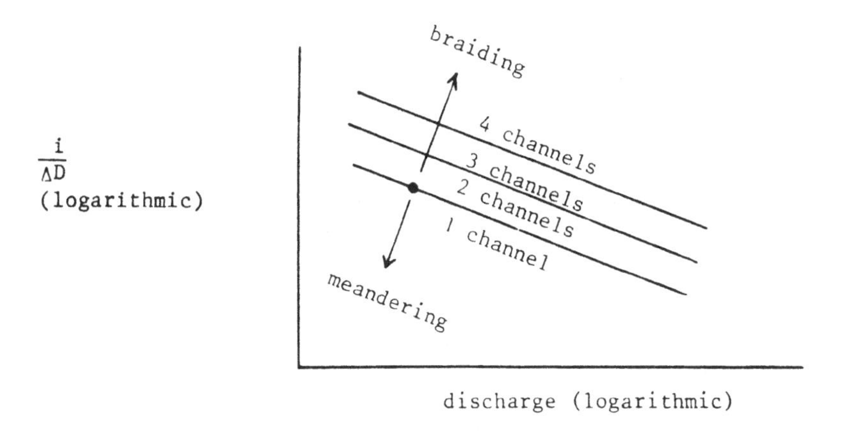

Figure 3. Channel pattern classification.

2.3. Resulting Scaling Procedure

Based on the observations made in Section 2.2, a scaling procedure can be derived that will result in a correct reproduction of the channel pattern. From Equation 4 it follows that:

$$n_{i/\Delta D} = (n_Q)^{\beta} \tag{5}$$

The relative density and the particle diameter D of the model are selected beforehand (see also below). Thus for a given slope i_m (and thus n_i) the value of n_Q can be obtained from Equation 5 (of course the only problem is the value of ß; see for this aspect Section 4). From the prototype value of the total discharge and from n_Q the total discharge in the model Q_m is obtained.

So applying the condition that the channel pattern (number of braids) in prototype and model should be similar, results (for given D_m, Δ_m and i_m) directly into Q_m. Given the total discharge in the model, all characteristics of the channels in the model are determined because these are self-forming regime channels. So the length scale, the depth scale, the distortion and all other scales cannot be influenced anymore! Most probably this will lead to scale effects. These effects are discussed in more detail in Section 3.

3. SCALE EFFECTS

In this Section it is attempted to estimate beforehand to what extent deviations from scale conditions will occur when a scale model is used, designed to reproduce the channel pattern correctly. It is quite obvious that the deviations are related to the degree of scaling down of the prototype. The larger the model, the smaller will the scale effects be. In designing such a model there are, however, a number of practical limitations. For the present model investigation these limitations are listed below:

- model dimensions as a maximum at about 50 x 20 m^2 because of costs, but also because of the required celerity of the model;
- model material sand, because light weight material (that would be preferable) is too expensive in large quantities;
- model material size should be about 200 μm (readily available within Delft Hydraulics), and maximum horizontal celerities according to Hickin and Nanson (1983).

3.1. Distortion

The channel characteristics in the model are completely determined by the model discharge. To evaluate to what extent deviations from scale conditions may be expected, it is necessary to be able to estimate these channel characteristics before-hand. In first approximation the Lacey equations for self-forming channels can be used. According to Lacey (1958) the average width B and the average depth $\overline{h}$ of self-forming channels are related to the discharge according to:

$$B \;::\; Q^{1/2} \qquad (6)$$

$$\overline{h} \;::\; Q^{1/3} \qquad (7)$$

Accepting these equations for the time being as a basis for the evaluation intended here, the width and depth scale are related to the discharge scale via:

$$n_B = n_Q^{1/2} \qquad (8)$$

$$n_{\overline{h}} = n_Q^{1/3} \qquad (9)$$

An estimate of the model distortion r is obtained via

$$r = \frac{n_B}{n_{\overline{h}}} = n_Q^{1/6} \qquad (10)$$

The above implies that for a width scale of $n_B = 10^3$ (preferably considering the available space and other reasons), the following scales are approximately found for the other scales:

$$n_Q = 10^6 \quad (Q_m = 100 \ \frac{1}{s})$$

$$n_{\bar{h}} = 10^2 \quad (h_m = 0.08 \ m \ during \ floods)$$

$$r = 10$$

So this scaling procedure results in a considerably distorted model. (It should be noted here that first experience from the model shows that the water depth is smaller than estimated here, resulting in a slightly less distorted model).

3.2. Froude Number

For a model discharge of about 100 l/s a fair reproduction of the channel pattern is probably obtained for a model slope of about 8.10^{-3} (see Section 4). Assuming the Chézy coefficient C_m in the model to be about 25 $m^{1/2}$/s, the average velocity in the channels is estimated to be about 0.4 m/s. This implies that the Froude number in the model is on the average in the order of 0.5 while in the prototype Fr = 0.2. Locally in the model even higher values of the Froude number may be found. Such high Froude numbers will certainly affect the flow pattern reproduction (see Struiksma, 1981) and consequently induce scale effects. For this reason in addition to the present scale model a
two-dimensional (depth-average) mathematical model will be used to determine the flow field (given a certain bed topography with channels and bars). The effect of the deviation of the Froude condition on the channel patterns cannot be evaluated beforehand, but for the time being this is assumed to be limited.

3.3. Roughness Condition

According to the roughness condition, that reads as:

$$n_C^2 = \frac{n_L}{n_{\bar{h}}} \ (= r) \tag{11}$$

the model distortion r should correspond to n^{2C}. Accepting $C_m = 25m^{1/2}$/s as a first estimate (resulting in $n_C = 3.5$), it follows that the roughness condition is reasonably well fulfilled. For rivers with smaller Chézy coefficients during floods this would be less favorable.

3.4. Bed Topography

As shown by Struiksma et al (1985) an important parameter for the reproduction of the bed topography in curved alluvial channels is the parameter λ_s/λ_w, where λ_s and λ_w are defined via:

$$\lambda_s = \frac{1}{\pi^2} h \left[\frac{B}{h}\right]^2 A\sqrt{\theta} \tag{12}$$

$$\lambda_w = C^2 h/2g \tag{13}$$

where A = empirical constant. The scale factor of λ_s/λ_w is derived as:

$$n_{\lambda_s/\lambda_w} = n_C^{-2} \cdot n_B^2 \cdot n_h^2 \cdot n_\theta^{1/2} \cdot n_A \tag{14}$$

According to the present experience of Delft Hydraulics $n_A = 0.5$. Introducing the other scale factors in Equation 14, results in $n_{\lambda_s/\lambda_w} = 5$. This deviation is mainly due to the model.

Consequently it may be expected that scale effects will be present in the reproduction of the bed topography which in turn may result in deviations in the reproduction of the channel characteristics (notably the point bar dimensions).

It should be noted that the value of n_{λ_s/λ_w} is rather high. This may indicate that the model channel dimensions are not estimated correctly here.

3.5. Curvature of Channels

The reproduction of the curvatures of the channels can also be evaluated by applying relationships from geomorphological sources (Leopold et al, 1964). From these the following scale rules can be derived:

$$n_{\text{meander length}} = n_B^{1.01} \tag{15}$$

$$n_{\text{radius of curvature}} = n_{\text{meander length}}^{1.02} \tag{16}$$

which would result in the following scale rule for the ratio radius of curvature/channel width:

$$n_{\text{radius of curvature}/B} = n_B^{.03} \tag{17}$$

which would result in a value of 1.23 for the present example.

Summarizing the above considerations, it should be stated that a model scaled according to the procedure outlined in Section 2.3 is subject to considerable scale effects. Therefore such a model should be used for comparative research only.

4. ADDITIONAL REMARKS

The following additional remarks can be made:

4.1. Selection of Model Slope and Discharge

The main problem in designing a model according to the procedure outlined in Section 2, is the fact that the Equation 4 is not completely known. This implies that the process of obtaining values for i_m and Q_m (given D_m and Δ_m), that result in a fair reproduction of the channel pattern in the Jamuna River, necessarily has to involve trial and error. As it is very cumbersome and time-consuming to change the slope of a model with approximate dimensions of 50 x 20 m^2, for the time being the slope of the model has been selected to be about 8. 10^{-3}. This slope was determined from an analysis of all data on channel patterns available in literature, after having selected a model discharge of about 100 l/s as desirable. At present tests are carried out with different model discharges and the resulting channel patterns are determined with aerial photography. Next braiding parameters are estimated (see Howard et al, 1970) and the model discharge (and thus n_Q) will be selected on the basis of the comparison of these characteristics with the characteristics of the Jamuna River itself.

4.2. Varying Discharge

In the preceding sections the model discharge was discussed as if it should have a constant value. From preliminary observations in a model, that has been designed according to the scaling procedure discussed before (see Figure 4), it appears that a succession of high and low flows results in a better reproduction.

4.3. Lightweight Material

If, indeed, Equation 4 is preferable over Equation 1, it follows that a less serious deviation from the Froude scale would result if light weight material could be used. This stresses the need for not too expensive light-weight material!

4.4. Testing of Similarity

In the end of 1987 the results of the model tests at Delft Hydraulics can be used to evaluate the possibilities of a model, scaled according to the procedure outlined above, to simulate prototype conditions.

Figure 4. Overview of movable bed model of the braiding Jamuna River during first tests.

5. ACKNOWLEDGEMENTS

The scale model investigation discussed here is part of the Jamuna Bridge Appraisal Study Phase II, which is funded by UNDP, the World Bank being the executive agency. The study is carried out by the combination RPT (Rendel, Palmer and Tritton, U.K.)/ NEDECO (Netherlands Engineering Consultants)/BCL (Bangladesh Consultants Ltd.); the JMBA (Jamuna Multipurpose Bridge Authority, Bangladesh) is the client. Delft Hydraulics is one of the partners of NEDECO for this study.

6. REFERENCES

Ferguson, R. T. (1984), 'The threshold between meandering and braiding,' Proc. 1st Conf. on Channels and Channel Control Structures, Southampton, U.K.

Hickin, E. J. and Nanson, G. J. (1983), 'Lateral migration rates of river bends,' *J. of Hydraulic Engineering*, **110**, pp. 1557–1567.

Howard, A. D., Keeth, M. E. and Vincent, C. L. (1970), 'Topological and geometrical properties of braided streams,' *Water Resources Research*, **6**, No. 6, 1674–1688.

Klaassen, G. J., Vermeer, K. and Uddin, N. (1988), 'Sedimentological processes in the Jamuna (Lower Brahmaputra) River, Bangladesh,' Submitted to *Fluvial Hydraulics* 1988, Budapest, Hungary.

Lacey, G. (1958), 'Flow in alluvial channels with sandy mobile bed,' *Proc. Instr. Civ. Engineers*, **9**, pp. 145–164.

Leopold, L. B., Wolman, M. G. and Miller, J. P. (1964), *Fluvial processes in geomorphology*, San Francisco, W. H. Freeman and Co.

RPT/NEDECO/BCL (1987), *Jamuna Bridge Appraisal Study Phase I*, Final report.

Struiksma, N. (1985), 'Prediction of 2-d bed topography in rivers,' *Journal of Hydraulic Engineering ASCE*, **111**, No. 8, 1169–1182.

Struiksma, N. (1981), *Recent developments in design of river scale models with mobile bed*, IAHR Symp. on River Engineering, Belgrade, Yugoslavia.

Struiksma, N., Olesen, K., Flokstra, C. and de Vriend, H. J. (1985), 'Bed deformation in curved alluvial channels,' *Journal of Hydraulic Research*, **23**, No. 1, pp. 57–79.

Struiksma, N. and Klaassen, G. J. (1988), *On the threshold between meandering and braiding*. Paper submitted for River Regime Conference, Wallingford, U. K.

Wijbenga, J. H. A. (1985), *Determination of river flow pattern on curvilinear coordinates*, Proc. 21st IAHR Conf., Melbourne, Australia.

MODEL SCALES FOR SAND BED CHANNELS

BOMMANNA G. KRISHNAPPAN
Rivers Research Branch
National Water Research Institute
Canada Centre for Inland Waters
Burlington, Ontario, L7R 4A6
Canada

ABSTRACT. Scale relationships for physical modelling of sand bed river channels are derived using friction factor equations that take into account the form drag caused by sand waves such as dunes and anti-dunes. An interesting aspect of the present derivation is that the friction factor equations developed by different investigators using different set of characteristics parameters result in identical scaling laws for dune regime and two out of three equations adopted for the present derivation show that the scaling laws applicable to dune regime are also valid for anti-dune regime.

1. SCOPE

For non-uniform and mobile boundary channel flows, Yalin (1971) had derived scale relationships of a distorted model by considering the following five basic criteria:

$$\lambda_{x_1} = 1; \lambda_{x_2} = 1; \lambda_S = n; \lambda_E = n; \lambda_{Fr} = 1 \tag{1}$$

where λ stands for the ratio of model value of a property to the prototype value of the same property and the dimensionless numbers, x_1, x_2 and Fr are defined as follows:

$$x_1 = \frac{V_* D}{\nu} \quad \text{(Shear Reynolds Number)}$$

$$x_2 = \frac{\rho V_*^2}{\gamma_S D} \quad \text{(Mobility Number)} \tag{2}$$

H. W. Shen (ed.), Movable Bed Physical Models, 73–79.

$$Fr = \frac{V}{\sqrt{gh}} \qquad \text{(Froude Number)}$$

The meaning of symbols appearing in relations (1) and (2) above are given below:

S = slope of channel bed
E = slope of the energy-grade-line
n = distortion = λ_y/λ_x
V_* = shear velocity

D = grain size
ν = kinematic viscosity of fluid
ρ = density of fluid
γ_s = submerged specific weight of sediment
V = average flow velocity
h = average depth
g = acceleration due to gravity
$x\&y$ = horizontal and vertical dimensions respectively

The scale relationships of Yalin are:

$$\lambda_D = (\lambda_y n)^{-1/2}$$

$$\lambda_{y_s} = (\lambda_y n)^{3/2}$$

$$n^{1/2}\left[1 - \frac{\ln\left(\frac{\lambda_D}{\lambda_y}\right)}{\ln\left(11\,\frac{h'}{D'}\right)}\right] = 1 \qquad (3)$$

and $\lambda_V = \lambda_y^{1/2}$

where h′ and D′ are prototype value of flow depth and grain size, respectively.

In deriving the scale relationships, Yalin had employed the friction factor equation corresponding to flows over plain bed and noted that a model designed according to the above scale relationships would not reproduce the correct water surface profile if the prototype flow consists of bed undulations such as ripples and dunes. He further showed that if the prototype bed consists of ripples, then the model would appear to be rougher than what it should be and the flow depth in the model would be larger than the required depth. The model response would be just the opposite if the prototype bed forms are dunes.

In this paper, Yalin's method has been extended to include flows with bed forms. A friction factor equation which considers both the skin friction and the form drag of the bed forms is used to derive a new scale relationship that could be used in place of the third relationship in equation (3). The details of the derivation of this new scale relationship and its application to model flows with different bed form regimes are discussed in what follows.

2. DERIVATION OF THE NEW SCALE RELATIONSHIP

A general expression for the slope of the energy grade line of mobile boundary flows with bed forms has been derived by Krishnappan (1985) as:

$$E = C \cdot \left(\frac{\gamma}{\gamma_s}\right)^L \cdot \left(\frac{R}{D}\right)^M \cdot \left(\frac{V^2}{gR}\right)^N \tag{4}$$

where C, L, M and N are given by:

$$C = 6.82^{6/(1-4\,a_1)} \cdot c_1^{\,4/(1-4\,a_1)}$$

$$L = \frac{4-4a_1}{4a_1 - 1} \tag{5}$$

$$M = \frac{3 - 4a_1 - 4b_1}{4a_1 - 1}$$

$$N = \frac{3}{4a_1 - 1}$$

The quantities a_1, b_1, and c_1 appearing in (5) are parameters of the following power relationship among the Mobility Number Y′ formed using the shear stress pertaining to the stein roughness, the total Mobility Number Y and the relative hydraulic radius (R/D).

$$Y' = c_1\, Y^{a_1} \left(\frac{R}{D}\right)^{b_1} \tag{6}$$

Knowing the actual relationship among Y′, Y and (R/D) in the above form, the values of a_1, b_1, and c_1 can be established for a particular type of bed form. Knowing a_1, b_1, and c_1, the slope of the energy grade line can be fully defined using equations (4) and (5).

Some of the existing friction factor equations of mobile boundary flows were expressed in the form of equation (4) for different bed form configurations and the values of C, L, M, and N are summarized in Table 1.

Table 1. Values of C, L, M, and N for different friction factor equations.

Name of Formula	Type of Bed Form	C	L	M	N	$\left(\frac{3(L-M)}{(M-3L-2)}\right)$
Engelund (1966)	Dunes	0.326	-4/7	-5/7	3/7	-3/7
	Antidunes	0.022	0	-1/3	1	-3/7
Garde and Ranga Raju (1966)	Ripples and dunes	0.098	0	-1/3	1	-3/7
	Antidunes	0.028	0	-1/3	1	-3/7
Kishi and Kuroki (1974)	Dunes	.0052	2	1	3	-3/7
	Antidunes	.0021	2	1/5	3	-27/39

Using equation (4), the scale relationship for the slope of the energy grade line can be evaluated as:

$$\lambda_E = \left[\frac{\lambda_\gamma}{\lambda_{\gamma_s}}\right]^L \cdot \left[\frac{\lambda_y}{\lambda_D}\right]^M \cdot (\lambda_{Fr})^{2N} \tag{7}$$

With the basic requirements that $\lambda_E = n$, $\lambda_{Fr} = 1$, and $\lambda_\gamma = 1$, and using the first two scale relationships of equation (3), equation (7) can be rearranged to get a relationship between distortion n and the vertical scale λ_y as:

$$n = \lambda_y^{3(L-M)/(M-3L-2)} \tag{8}$$

3. DISCUSSION OF RESULTS

It is interesting to note that the exponent of λ_y in equation (8) takes a value of (-3/7) for all three friction factor formulae listed in Table 1 in the case of dunes and for two out of three formulae in the case of antidunes (see the last column in Table 1). It is also surprising that three different equations with different exponents of the governing parameters produce the same scale relationship, namely.

$$n = \lambda_y^{-3/7} \tag{9}$$

for modelling flows with dune covered beds. All three equations are of empirical nature and are based on different approaches. For example, Engelund had expressed Y' as a function of Y alone without any consideration of (R/D) parameter whereas Kishi and Kuroki expressed Y' as a function of both Y and (R/D). Garde and Ranga Raju assumed a Manning type equation and determined the coefficient and exponents using different data sets from the ones used by Engelund and Kishi and Kuroki. Under these circumstances, it is a remarkable coincidence that all three equations give the same scaling law.

For the case of flows with antidunes, two out of three equations indicate that the modelling laws applicable to dune regime are also applicable to antidune regime. Only Kishi and Kuroki's equation gives two different scaling laws for dunes and antidunes. In the light of this result, the formulation of Kishi and Kuroki's equation for antidune regime should be reexamined. It is highly desirable, from a practical point of view, to have the same modelling laws apply to both types of bed forms.

The modelling law given by equation (9) is also practicable as it yields model distortions that are not too excessive. Table 2 gives the values of distortion n for a range of vertical scale of the model. Figure 1 shows a plot between n and λ_y.

Table 2. Values of distortion n for various values of λ_y.

For λ_y =	1/100	1/50	1/30	1/25	1/20	1/15 /10	1
n =	7.2	5.3	4.3	4.0	3.6	3.2	2.7

4. SUGGESTIONS FOR DISCUSSION

In discussing the scale relationships shown as equation (3), Yalin showed that a model that reproduces the energy loss due to skin friction correctly is incapable of yielding correct losses due to form drag. The new model scale derived in this paper ensures correct total energy losses due to both skin friction and form drag whereas the individual components may not be correctly modelled. Importance of modelling individual components of energy losses correctly could be a topic for further discussion.

5. REFERENCES

Krishnappan, B. G. 1985. Modelling of unsteady flows in alluvial streams. *Journal of Hydraulic Engineering*, ASCE, Vol III, No. 2, 257-266.

Yalin, M. S. 1971. *Theory of Hydraulic Models*. Macmillan Publishing Co.

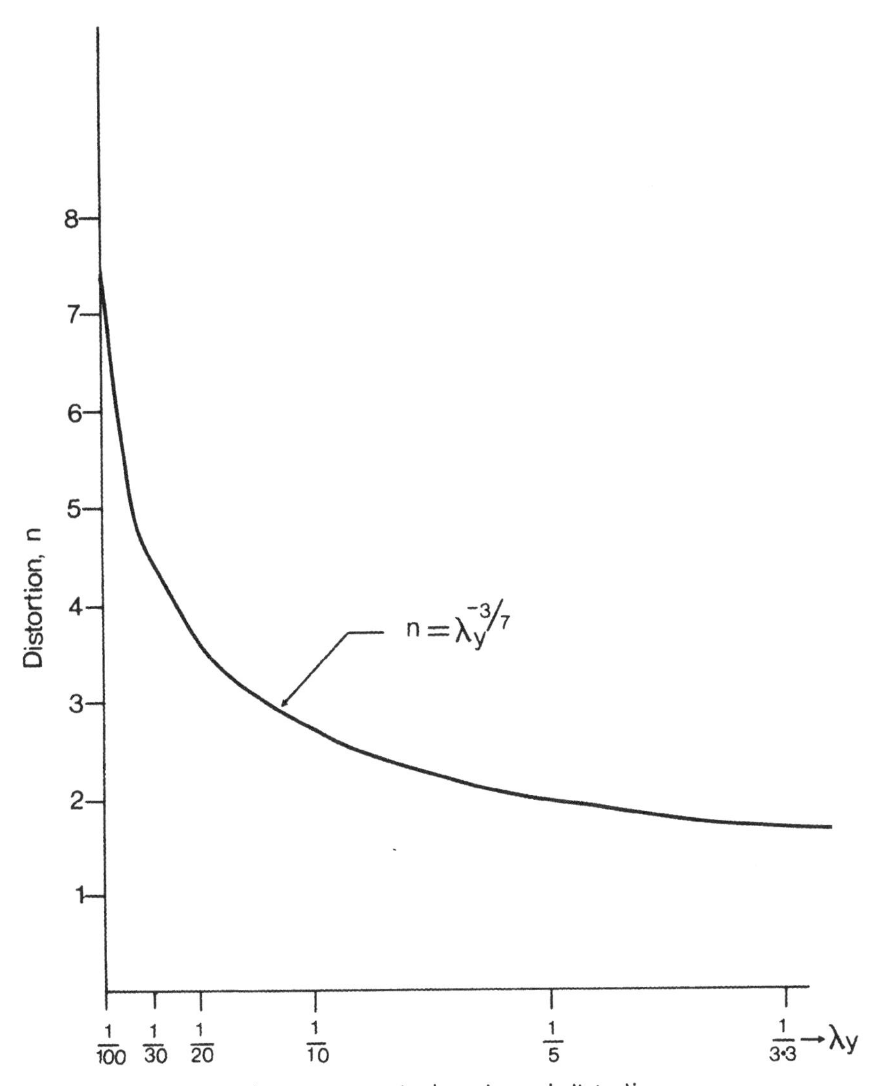

Fig.1 Relation between vertical scale and distortion for mobile boundary models.

MOVABLE MODELS FOR GRAVEL BED STREAMS

JOÃO SOROMENHO ROCHA
Hydrology and River Hydraulics Division
Laboratório Nacional de Engenharia Civil (LNEC)
Lisbon, Portugal

ABSTRACT. The main purpose is to analyse the modelling of gravel bed streams with movable bed models, comparing with same modelling of sand bed streams. The stream processes are the same for gravel and sand bed streams. The critical point is to develop a sound physical theory to be applied to basic stream processes, using simultaneously laboratory, numerical, analytical, and field observation efforts.

1. CHARACTERIZATION OF GRAVEL BED STREAMS

The assertion by Hey, Bathurst, and Thorne (1982) that there is a lack of knowledge regarding the channel processes is still valid. Although in recent years there has been an increase in research on gravel bed channels, the research emphasis in the past on research on rivers with finer alluvial beds greatly influences the current knowledge of movable bed channels. The above mentioned book and the presentations of the International Workshop on Engineering Problems in the Management of Gravel-Bed Rivers held at Gregynog, Newtown, U.K., in June 1980, and the differentiation between medium size bed material (sand) and gravel beds in this workshop signify that a unified theory of basic channel processes does not exist. To enable the development of sound engineering principles, modeling methods need to be analyzed to establish better guidelines for river management, and to identify future paths.

Three ways to examine gravel bed streams are: 1) sediment size distributions; 2) flow regime and sediment transport; and 3) stream processes. These are discussed below.

H. W. Shen (ed.), Movable Bed Physical Models, 81–90.

1.1 Sediment Size Distributions

Gravel is commonly defined by a sediment grain size distribution of 2mm-60mm. It is rare to find a natural stream bed with a distribution of grain sizes limited to that range. The gravel bed stream sediments usually encompass a wider range of grain sizes. A gravel bed exists if the bed composition is monomodal and the medium value is between 2mm and 60mm. There is no doubt that this designates a gravel bed stream. It would be useful to determine a minimum percentage of sediment bed material size falling between 2mm and 60mm in order to define a stream as a gravel bed stream. This minimum percentage could be 30% or 40%. Since, in general, gravel bed streams are separated from sand bed streams, the upper limit of gravel size may also consist of a significant amount of cobbles and even larger material.

The wide range of grain sizes present in a gravel bed river leads to important features of bed heterogeneity, such as surface armouring, vertical grain size variability, and bimodal features of bed composition. Two additional features which characterize the explicit division of movable bed streams into sand bed or gravel bed rivers will be discussed below.

Gravel bed rivers usually are armoured at their surface during periods when no bedload transport occurs. The minimum grain size which causes a bed composition to take on characteristics of an armoured bed is a fundamental question. Bray (1982) states that an armoured bed is generally characterized by a unimodal grain size distribution with size ranging from 16mm to 128mm, or gravel-cobbles size. Rivers in Portugal have similar values. However, armoured layers cannot be defined satisfactorily by designating a certain size range. It is necessary to search for the minimum grain size necessary to form an armoured layer. Bed material with D_{50} = 1.5mm and D_{90} = 8mm may still form an armoured layer (see Figure 1). Further research is needed in this area. The presence of a bimodal bed composition may accelerate the armouring process. A bimodal bed composition depends on geologic characteristics of the watershed.

1.2 Flow Regime and Sediment Transport

The discharge, sediment input, valley slope, and bed composition are usually considered to be variables (Hey, 1982). In fact, there is a continuous interaction between all these variables. It is difficult to identify which variables are dependent and which variables are independent.

The rivers adjust their overall shape and channel dimensions (dependent variables) and the measured data on rivers and laboratory canals, the process analysis, and the theoretical deductions made since the beginning of the 20th century admit the above four variables, i.e., discharge Q, sediment input Q_S, valley slope S_V, and bed composition D, as independent variables. Simultaneously, it is recognized that natural channels may be stable over geological time periods and in shorter time scales of less than 100 years. In fact there is a continuous interaction between all the variables in the system, due to the operation of feedback

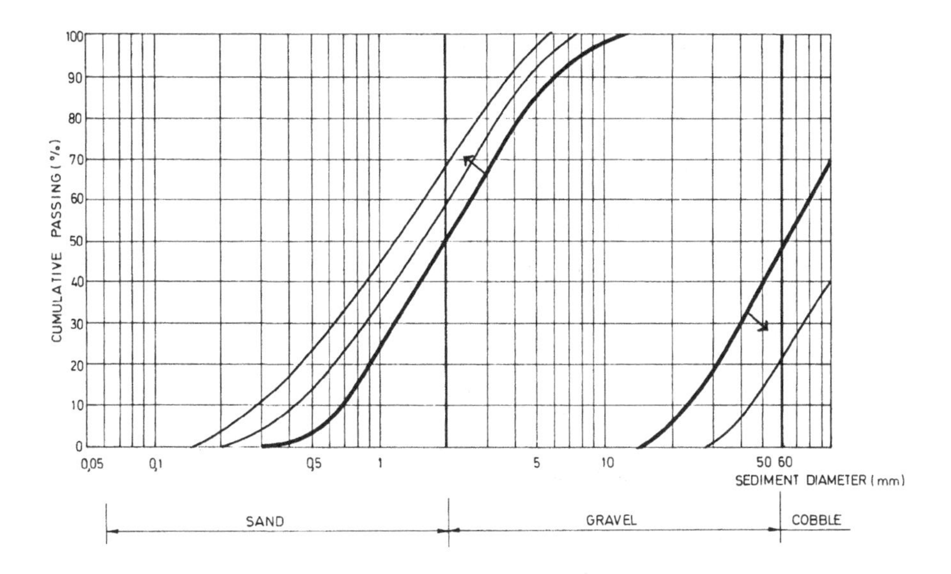

Figure 1. Bed Compositions.

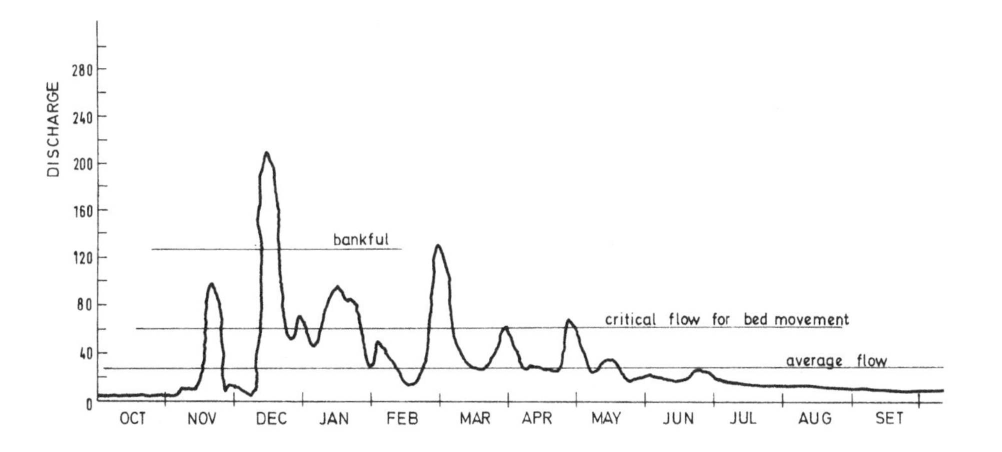

Figure 2. Flow Regime.

mechanisms, precluding the identification of the dependent and independent variables.

The flow regime or the time sequence of different values of discharges in temperate climates of Europe and North America is related to atmospheric events. The river beds are modified by these events. The precipitation events which cause the flow regime affect entire watershed areas. These areas have different geological, morphological, and natural soil cover characteristics. The present river beds are a consequence of these greater factors and are subject to very long-term cumulative effects of these interrelated factors (see Figure 2).

In Portugal, gravel beds exist in very large rivers and in first and second order streams that are subject to dry periods. One of the fundamental factors for explaining gravel bed occurrences is geology. This factor is not considered in river regime or sediment transport analysis. The sediment transport analysis has been studied in detail and several formulations have been developed. But as Parker (1982) stated, "the problem of gravel bedload seems to become progressively more exasperating with each successive investigation." The general conclusion is that existing theoretical analyses are far from providing solutions to engineering problems. It is important to conduct extensive basic research. Hopefully, the results from the Large Tilting Flume tests at LNEC may contribute to knowledge on gravel bed streams, which are important in Portugal.

1.3 Stream Processes

There are seven governing or process equations which link the two sets of variables: the independent variables - Q, Qs, D, Dr, Dl, S_v (Q = water discharge; Q_s = sediment discharge; D = sediment diameter; D_r = sediment diameter in right bank; DL = sediment diameter in left bank; S_v = valley slope); and the dependent variables - U, R, S, P, d, p, ϵ (R = hydraulic radius; S = energy slope; P = wetted perimeter; d = depth of water; p = sinuosity in plan; ϵ = morphological factor of bed). These seven equations are the equations of continuity, flow resistance, sediment transport, bank erosion, bar deposition, and meandering. For centuries, the flow resistance and the bed and banks shear stresses have been the most important problems to be solved. Today these problems are not theoretically precisely solved even for rigid boundary problems, such as that for spillways and other large hydraulic structures. Solutions for these problems for mobile beds are difficult to obtain and thus mobile bed models are used to obtain approximate solutions for engineering applications.

The attempt to link the bedform characteristics with stream processes are handicapped by a lack of knowledge of fluvial physics. For instance, Church and Jones (1982) presented a table with four bedform classifications (micro, meso, macro, and megaforms) associating spacial variables (D, d, B, λ) (D = sediment diameter; d = depth of water; B = width of channel; λ = meander length) and time scales (the time for a floodwave to pass through a reach). This idea has not received appropriate attention. The time influence of stream processes phenomena is an area that needs more attention. Even without knowing the reasons for the occurrence of different bed features, the theoretical solution for these phenomena requires a good physical insight into the effects of short-term transient floodwaves as well as the long-term cumulative effects of the flow variations on streambeds (Figure 3).

The theoretical solution might be presented by a very simple final formulation. It is difficult to obtain some simple equations in the form of the following:

$$\lambda = 2\pi B$$

as is found for meandering phenomena, in rivers and in other earth environments.

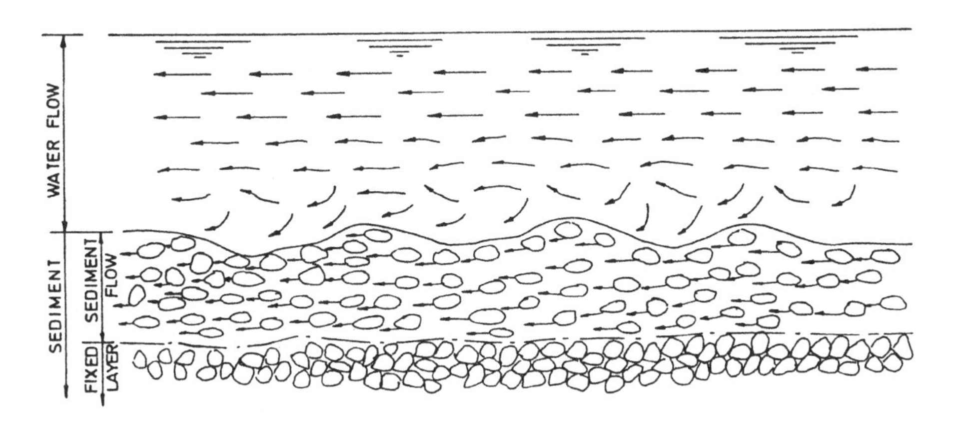

Figure 3. Bed Movements Schematization.

2. BASIC DIFFERENCES BETWEEN GRAVEL BED AND SAND BED STREAMS

Theoretical analyses of stream processes do not reveal significant differences between gravel bed and sand streams. The flow regime is not a good limiting indicator. The sediment transport analysis over armoured layers and the sediment size characteristics on the stream bed indicated that there are two different fluvial processes for gravel and sand bed streams. However, these differences do not necessarily require different theoretical approaches, except for the convenience of analysis. More research is needed in this area.

3. SIMILITUDE CRITERIA - APPLICATION TO GRAVEL BED STREAMS

Discussions on sediment transport phenomena and physical modeling were presented by the author (1986) at the IAHR Symposium, and the following additional comments can be added. Two important branches regarding similitude criteria, basic research and applied hydraulic engineering, must be developed. For basic research, the similitude criteria must be developed from a strong physical and fundamental approach. This may require combined efforts from an interdisciplinary team which consists of laboratory experts for experimental efforts, computer experts for advances in process analysis, field observation experts for sound validation purposes, as well as others. This would be a step toward finding better analytical solutions for movable bed streams. This is a task for research institutes and will be discussed in the next paragraph. For applied hydraulic engineering, the task is to search for immediate solutions to fluvial problems with limited field data by using international experience in physical modeling.

The application of similitude criteria for the probability of armouring on a gravel bed may not be theoretically different from the armouring of fine sediment sizes on a sand bed stream. If it is assumed that the two types of armouring are different, attention is focused on two particular problems: 1) stable armouring, in which the dominant problem is flow resistance, and 2) bed movements, in which the dominant problem is sediment transport without armouring. In a great number of gravel bed rivers, the stable armouring occurs during dry seasons and bed movements occur during flood events. In both cases, the analysis of flow regime and sediment transport are involved.

The formations of bedforms on gravel bed and on sand bed are different and require more theoretical analysis.

4. THE NEED TO DISTINGUISH SIMILITUDE CRITERIA FOR GRAVEL AND SAND BED STREAM

Similitude criteria developed during a long period of time by many researchers can be applied both to gravel or sand bed streams. As the author presented in 1981, 1984, and 1986, there are no reasons to search for different similitude criteria for gravel and sand bed streams.

As explained by Yalin (1986), the difference in solutions from applying different known similitude criteria to each case may be caused by the emphasis of different properties in each similitude criteria. The differences may not be significant if a particular set of grain size compositions is used (Figure 4).

5. DEVELOPMENT AND APPLICATION OF MOVABLE BED MODELS

There is a great difference in the development and the application of bed models for basic research problems or with applied engineering problems. The selection of test materials, the runs, and range of significant variables are all dependent on those two objectives.

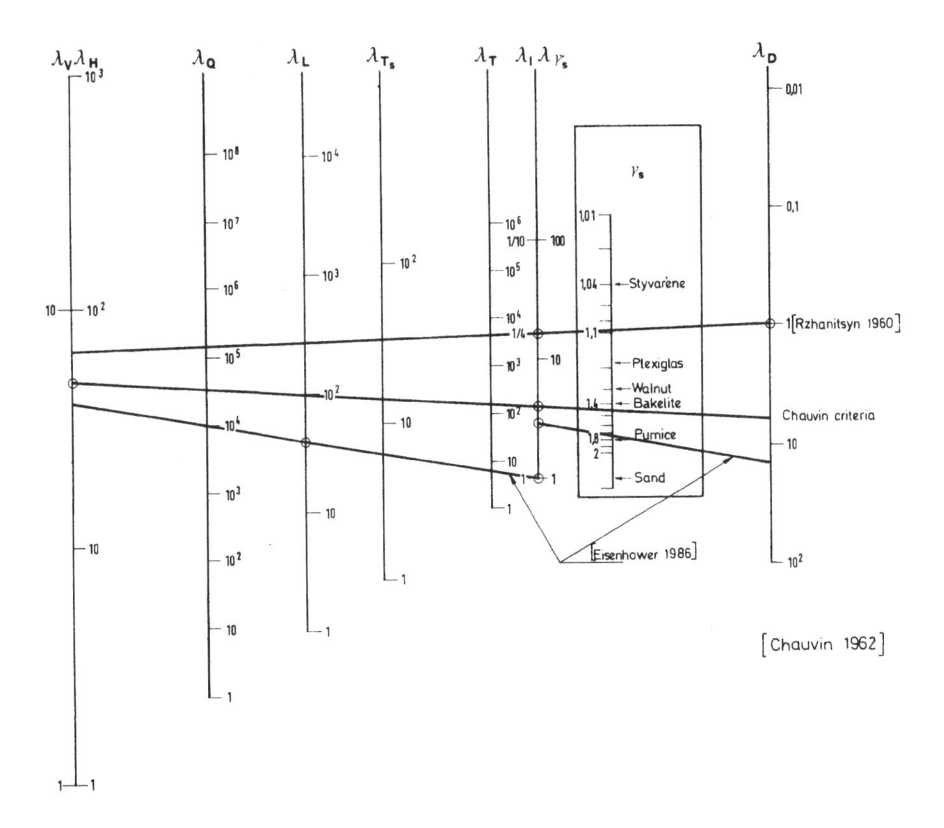

Figure 4. Similitude Criteria for Movable Bed Models.

5.1 Test Materials and Armouring

The study of hydraulic engineering problems are usually constrained by time and other restrictions. It may be difficult to secure several different materials unless the particular laboratory, such as at LNEC, has a large stock of natural sediments, bakelite, walnut shell, pumice, and plastic. It is the experience of the writer that it is not a critical problem if the similitude criteria and the existing material grain sizes do not permit a natural armouring on the physical model.

In the study of basic research problems, a particular material and its respective mobile bed size composition may be selected. Paradoxically, a great amount of laboratory experimental programs were dealing with almost uniform or narrow range bed compositions, for the convenience of analysis. In order to investigate armouring in canals, nonuniform sediment sizes must be used. In the designing of an experimental program, the author is not sure of the conditions to use a uniform or a natural bimodal sediment composition.

5.2 Test Runs and Time Scaling

Similarly, the test runs and time scaling aspects must be analyzed differently for a study case and an experimental program. For applied engineering problems, field data is usually scarce for the analysis of vertical bed movements, horizontal bed movements, and water movements. If sufficient field data are available at different time intervals, it is possible to calibrate model results with field data adequately. However, this is rare.

A basic research program to identify time scaling laws is one of the author's future tasks with results from the Large Tilting Flume of LNEC.

5.3 Steady and Unsteady Flows

Some observations may be made with both steady and unsteady flows. The author presented opinions (Rocha, 1986) about guidelines for using hydraulic models to study the development of river morphology. Currently it is much easier to run unsteady discharges in physical models. However, a good set of field data for unsteady flows is rather scarce.

For a basic research program, it is not necessary to emphasize the needs to conduct extensive research on unsteady discharge conditions.

6. SYNTHESIS

The gravel bed streams are fundamentally characterized by the occurrence of armoured layers, which is a major difference between gravel and sand bed streams.

The stream processes are the same for gravel and sand bed streams. The occurrence of gravel bed or sand bed streams is related to the geological characteristics of watersheds.

Yalin (1986) explained the possible differences of applying different similarity criteria to each case study.

For applied hydraulic engineering problems, model scales are verified by only a limited amount of field data. Time scales are usually a calibration factor. On the other hand, scientific research on sediment problems uses mostly similarity criteria theoretically derived by similarity theory. It is limited by incomplete knowledge of basic stream process, sediment motion, and turbulent characteristics.

The critical point is to develop a sound physical theory to be applied to basic stream processes, through laboratory investigations, numerical as well as analytical analysis, and field observations.

7. SUGGESTIONS FOR DISCUSSIONS

All of the previously mentioned considerations are suggestions for discussions, as are the following topics:

- The methods to incorporate geological parameters into stream process analysis;
- The increase of joint ventures for research programs including laboratory studies, analysis from both mathematical and physical perspectives, and field observations;
- The determination of the best similarity criteria to be applied to each case study.

8. REFERENCES

Hey, R. D., Bathurst, J. C., and Thorne, C. R. (Eds.), 1982. *Gravel-bed rivers. Fluvial processes, engineering and management.* New York: John Wiley and Sons.

Bray, D. I., 1982. Flow resistance in gravel-bed. In *Gravel-bed rivers. Fluvial processes, engineering and management* (R. D. Hey, J. C. Bathurst, & C. R. Thorne, Eds.). New York: John Wiley and Sons, 109-137.

Hey, R. D., 1982. Gravel-bed rivers: form and processes. In *Gravel-bed rivers. Fluvial processes, engineering and management* (R. D. Hey, J. C. Bathurst, & C. R. Thorne, Eds.). New York: John Wiley and Sons, 5-13.

Parker, G., 1982. Discussion of White and Day - Transport of graded gravel bed material. In *Gravel-bed rivers. Fluvial processes, engineering and management* (R. D. Hey, J. C. Bathurst, & C. R. Thorne, Eds.). New York: John Wiley and Sons, 214-220.

Church, M., and Jones, D., 1982. Channel bars in gravel-bed rivers. In *Gravel-bed rivers. Fluvial processes, engineering and management* (R. D. Hey, J. C. Bathurst, & C. R. Thorne, Eds.). New York: John Wiley and Sons, 291-338.

Rocha, J. S., 1981. Scaling laws on distorted models. *Proc. ISEH IAHR Congress,* New Delhi.

Rocha, J. S., 1984. Scales for distorted models with bends. River meandering. *Proc. Rivers '83 Conf.*, New Orleans, 974-984.

Rocha, J. S., 1986. Modeling of the development of river morphology. *Proc. IAHR Symposium 86, Toronto, Scale effects in modeling sediment transport phenomena*, 113-123.

Yalin, M. S., 1986. Some aspects of physical modeling of flows with a free surface. *Proc. IAHR Symposium 86, Toronto, Scale effects in modeling sediment transport phenomena*, 1-9.

SEDIMENT TRANSPORT UNDER NONEQUILIBRIUM CONDITIONS

SUBHASH C. JAIN
Iowa Institute of Hydraulic Research
The University of Iowa
Iowa City, Iowa 52242 U.S.A.

ABSTRACT. The concept of a spatial lag between the local sediment-transport rate and the equilibrium sediment-transport rate is described. Needs for improved scaling laws for sediment motion and interpretation of movable-bed model results are put forward. Suggestions for future research are presented.

1. INTRODUCTION

The solution of many problems in river hydraulics depends on results obtained from properly designed and operated physical models. An understanding of similarity criteria and the proper selection of model scales and material are important for the interpretation of model results. The established scaling laws for sediment motion are based on experimental results obtained under dynamic equilibrium conditions, while physical models are primarily used to study sediment transport under nonequilibrium conditions. The local sediment-transport rate under nonequilibrium conditions is different than the equlibrium sediment transport rate. Shoaling takes place if the former is larger, and scour occurs if the latter is larger.

Movable bed models based on the established scaling laws reproduce the overall prototype scour and fill patterns. There may exist a spatial lag between the locations of the shoaling and scour regions in the model and prototype. Figure 1 shows a comparison of model and prototype bed profiles for one section of the Absecon Inlet model (Jain and Kennedy, 1979). The reproduction of the bed profile would be satisfactory if the model profile were shifted about 500 ft (prototype), which corresponds to one foot in the model. The spatial lag between the model and prototype bed profiles is probably due to incorrect simulation of the spatial lag between the local sediment transport rate and the equilibrium sediment transport rate. There is a need for improved criteria for similarity between model and prototype.

H. W. Shen (ed.), Movable Bed Physical Models, 91–95.

2. SPATIAL LAG

The concept of the spatial lag can be best explained by considering the flow of water with a given flow depth and velocity over a movable bed. If the sediment-transport rate at the upstream boundary of the movable bed is less than the equilibrium sediment-transport rate, the local sediment transport rate will increase until it reaches the equilibrium value which is governed by the flow conditions and sediment properties. A certain distance, termed as a spatial lag, is required to entrain enough sediment by the flow to reach at its equilibrium rate. The existence of the spatial lag can be explained as follows. The sediment transport process can be considered to consist of two separate processes: entrainment and deposition of sediment. As the water flows over the movable bed, it entrains sediment from the bed, a portion of the already entrained sediment is deposited at the bed, and the remaining entrained sediment is transported downstream. These processes are shown schematically in Figure 2. The entrainment rate (sediment mass/unit time/unit area) depends on flow conditions and sediment properties, and should be constant as long as the latter are constants. However, the deposition rate (sediment mass/unit time/unit area) depends on the concentration of the sediment transported and will increase along the bed until it becomes equal to the entrainment rate at some downstream section. The sediment discharge at this section is equal to its equilibrium value, and is given by the shaded area in Figure 2. The equilibrium sediment transport rate per unit width, G^*, can be expressed as:

$$G^* = G + \int_0^L (E - D)dx \tag{1}$$

where E = sediment-entrainment rate per unit area, D = sediment-deposition rate per unit area, L = spatial lag, G = local sediment-transport rate at x = 0, and x = longitudinal distance. Assuming:

$$\frac{D}{E} = f\left(\frac{x}{L}\right)$$

equation (1) can be written as:

$$G^* - G = EL\int_0^1 \left[1 - f\left(\frac{x}{L}\right)\right] d\left(\frac{x}{L}\right) \tag{2}$$

The spatial lag from Equation (2) is given by:

$$L = \frac{G^* - G}{\alpha E} \tag{3}$$

where:

$$\alpha = \int_0^1 \left[1 - f\left(\frac{x}{L}\right)\right] \, d\left(\frac{x}{L}\right) \tag{4}$$

Clearly, the spatial lag is zero for $G = G^*$. A similar explanation for the existence of the spatial lag can be given for the flow condition where the local sediment discharge is larger than its equilibrium value.

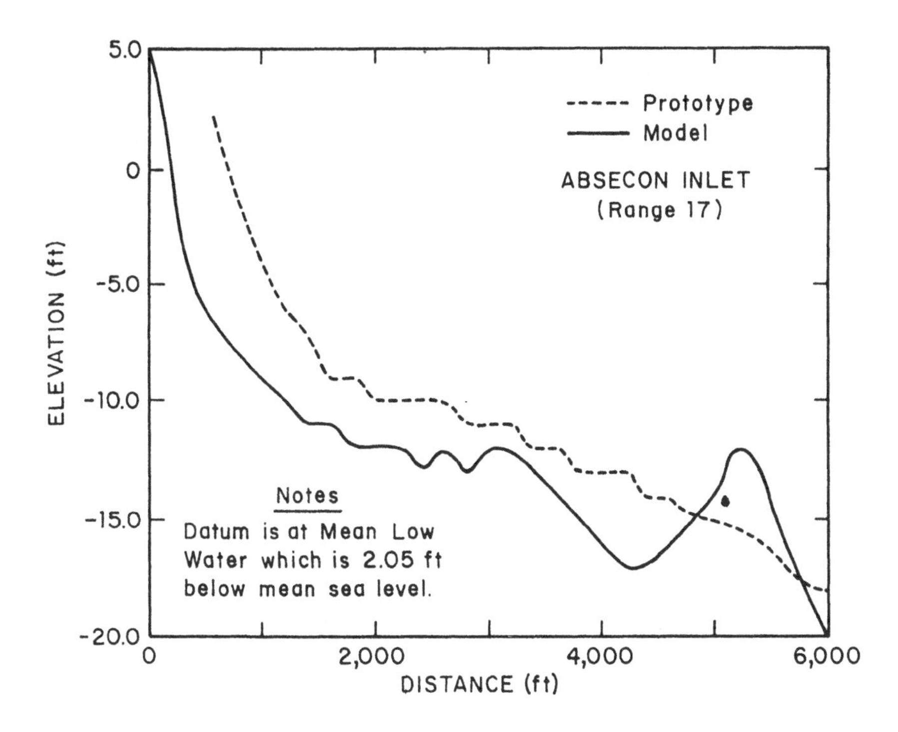

Figure 1. Comparison of model and prototype beach profiles at range 17 of the Absecon Inlet model (Jain and Kennedy, 1979).

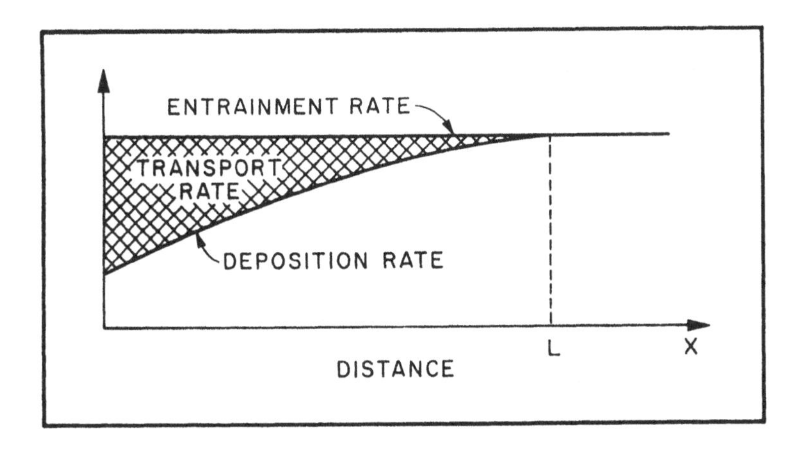

Figure 2. Variation of entrainment and deposition rates.

Some of the recent mathematical models (Holly, et al., 1987) for simulation of bed evolution in mobile-bed rivers proposed a bedload-loading law of the form:

$$\frac{dG}{dx} = a(G - G^*) \tag{5}$$

in which a = coefficient to account for the spatial lag between G and G*. Equation 5 is based on the experimental results of Bell and Sutherland (1983) presented in Figure 3. According to these results, the coefficient a is a function of time, which is probably due to the formation of a scour hole in the bed in their experiments. The experimental results of Mosconi and Jain (1986) also support the existence of the spatial lag. At present there is no predictor available to evaluate the coefficient a in Equation 5.

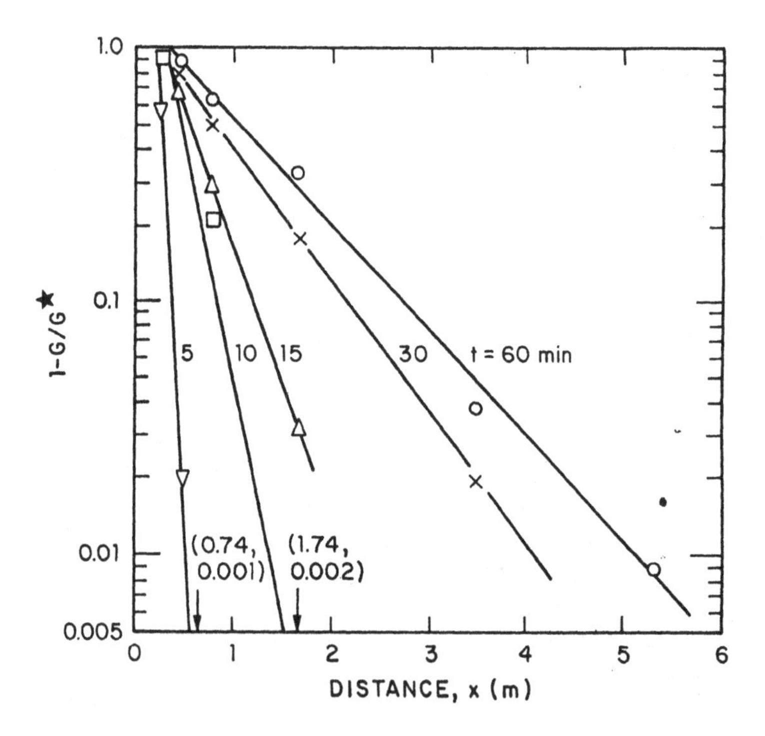

Figure 3. Spatial variation of the ratio local transport rate to local equilibrium capacity for different times; water discharge = 0.159 m³/s (Bell and Sutherland, 1983).

3. SUGGESTED RESEARCH APPROACH

The improvement of knowledge of entrainment and deposition processes will lead to a better understanding of the spatial lag, and will require undertaking fundamental studies on particle and turbulent fluid interaction. Recent experimental studies on particle motions near the bottom in turbulent flow in an open channel (Sumer and Diegaard, 1981) have shown that the mechanism of particle motion close to the wall can be explained by the model of the bursting process in turbulent boundary layers. These studies examined the behavior of a single particle. For a large number of particles in motion, the interaction between particles becomes important. These interactions may be direct or indirect. The direct interaction refers to the collision between particles, while the modification of the fluid-velocity field in the interspace between particles is referred to as indirect interaction. It is conjectured at the present time that the increase with distance of the deposition rate of the entrained sediment depicted in Figure 2 is due to the interaction effects of the particles. Experiments, cleverly conceived and designed and interpreted with skill, are necessary to establish scaling laws for the spatial lag.

4. REFERENCES

Bell, R. G., and Sutherland, A. J., March 1983. *Nonequilibrium Bed Load Transport by Steady Flows.* Journal of Hydraulic Engineering, ASCE, 109(3).

Holly, F. M., Jr., Cunge, J. A., Wignyouskarto, B., and Einhellig, R., August 1987. *Coupled Implicit Simulation of Movable-Bed Rivers.* Proc. National Conference on Hydraulic Engineering, ASCE, Williamsburg, VA.

Jain, S. C., and Kennedy, J. F., February 1979. *An Evaluation of Movable-Bed Tidal Inlet Models.* GITI Report 17, U.S. Army Coastal Engineering Research Center, Fort Belvoir, VA.

Mosconi, C. E., and Jain, S. C., March 1986. *Evolution of Armor Layer.* Third Int. Symp. on River Sedimentation, Jackson, MS.

Sumer, B. M., and Deigaard, R., August 1981. *Particle Motions Near the Bottom in Turbulent Flow in an Open Channel, Part 2.* Journal of Fluid Mechanics, 109.

ASPECTS OF MODELLING SUSPENDED SEDIMENT TRANSPORT IN NON-UNIFORM FLOWS

BERND WESTRICH
Universität Stuttgart
Institut für Wasserbau
Pfaffenwaldring 61
D-7000 Stuttgart/Vaihingen
Federal Republic of Germany

1. GENERAL REMARKS

The following consideration refers to non-equilibrium sediment transport problems in alluvial rivers caused by non-uniform flow situations. Since the existing modelling criteria are derived from uniform equilibrium transport in an alluvial bed consisting of particles with a diameter larger than about 0.2mm, one has to be careful in physical modelling of erosion and sedimentation processes. Selected problems related to the sedimentation of non-cohesive, very fine sediment (wash load) in river reservoirs are discussed in the following.

2. TRANSPORT CAPACITY APPROACH

River flows with almost straight alignment and regular cross section can be considered quasi-uniform in terms of suspended sediment transport as the adaptation length (required to establish the vertical equilibrium concentration profile) for erosion ($L_{A,E}$) or sedimentation ($L_{A,S}$) is much less than the backwater section:

$$L_{A,S} \sim L_{A,E} << L_{backw} \tag{1}$$

H. W. Shen (ed.), Movable Bed Physical Models, 97–105.

Therefore, local equilibrium transport can be approximately assumed. There are two backwater flow sections with different transport and sedimentation conditions:

Upstream section: sedimentation rate > erosion rate
Downstream section: pure sedimentation; no erosion

Sedimentation starts as soon as the limiting concentration C_T and the critical sedimentation bed shear

$\tau_{crit,s}$,

respectively, has been reached. Transport and sedimentation can be correlated to the transport capacity parameter as follows (Itakura, Kishi, 1980; Westrich, Juraschek, 1985):

$$C_T = k \frac{Q}{(Q_s - Q)} \; \frac{u^2}{gh} \; \frac{u_*}{v_s} \; \frac{u_*}{u} \tag{2}$$

$$k = f\left(\frac{Q}{(Q_s - Q)} \; \frac{u_*^2}{gd} \; \frac{u_*}{v_s} \right) \tag{3}$$

where C_T is the maximum possible depth averaged sediment concentration in parts per volume to be transported without deposition, h the flow depth, v_s the sediment fall velocity, u the flow velocity, and u_* the bed shear velocity.

The transport efficiency coefficient k depends on the deformation and the erodibility of the bed material that is the sediment Froude-number Fr_* and the fall velocity parameter u_*/v_s l(Wiuff, 1985). The value of k is lowest for a flat river bed with non-erodible large grains (Figure 1.A), it is higher for immobile bed forms (Figure 1.B) (ripples or dunes) and erodible flat bed (Figure 1.C), and highest for fully developed equilibrium transport with migrating bed forms (Figure 1.D). The interaction between the suspended sediment and the bed forms is strongest for fully developed dunes as they produce large scale turbulent eddies which are most effective for the sediment transport. From relation (3), the criteria for the model scales can be derived as proposed by H. W. Shen (1985).

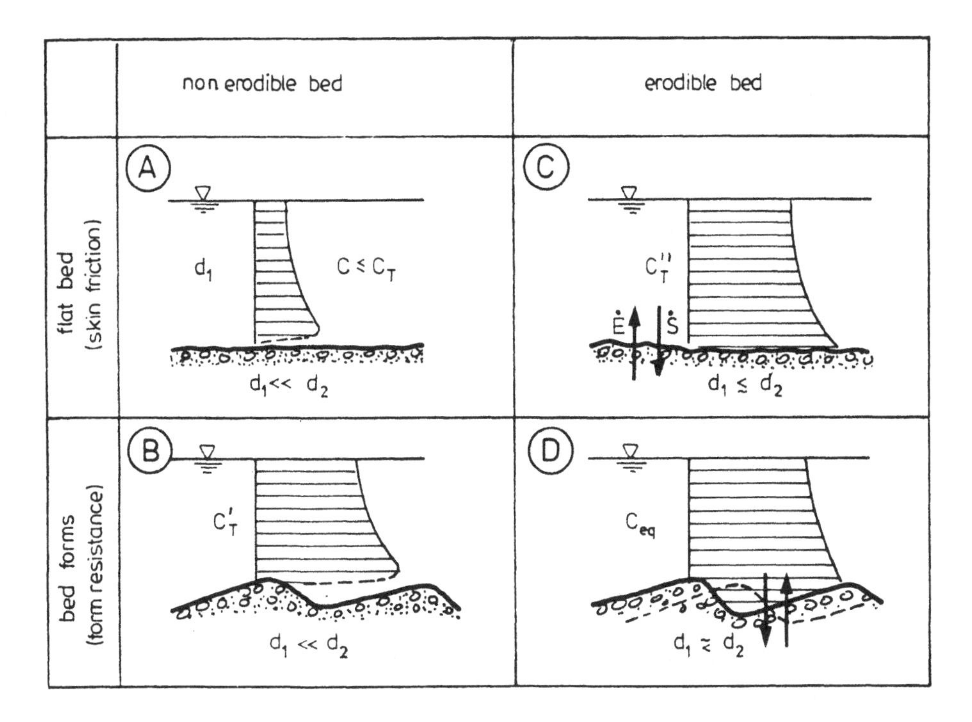

Figure 1. Different types of suspended sediment transport.

3. BACKWATER FLOWS (RIVER RESERVOIRS)

3.1 Regular Flow Cross-Section

Upstream of a river dam there is a continuous reduction of the transport capacity with a corresponding decrease of the sediment concentration. In the section where erosion and sedimentation are acting simultaneously, the bed features have to be reproduced to establish the longitudinal development of the deposition and, hence the time scale of the sedimentation process. Even if the bed load is relatively small, its influence on the turbulence production due to bed forms may be important. Beyond the flow cross section with critical erosion shear ($\tau < \tau_{crit,E}$) there is only sedimentation without any interaction at the river bed and, hence without bed form effects. Only the skin friction coefficient ($\lambda = \lambda'$) and the particle settling are relevant. Therefore, the length of the far downstream section with pure sedimentation depends on the velocity ratio u_*/v_s and λ.

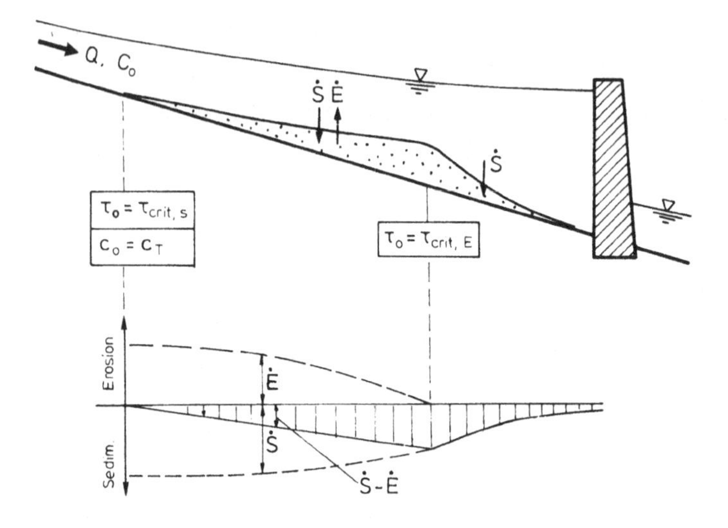

Figure 2. Longitudinal sedimentation in backwater flow.

Due to the gradually varied backwater flow, selective sedimentation takes place. That is, suspended particles settle according to the longitudinal development of the flow. This phenomena requires the reproduction of the sediment grading curve that is for the suspension:

$$\left(\frac{v_s}{v_{s50}}\right)_r = 1 \tag{4}$$

3.2 Compound Flow Cross-Sections

Many river reservoirs have a main channel and flat inundated plains. The transport capacity strongly varies across the river resulting in a strong deposition at the near bank strip of the flood plain. Usually the gradient of the sediment transport capacity is much larger in the lateral direction than in the longitudinal one. The lateral velocity distribution is important to ensure the similarity of the interfacial shear and the associated lateral dispersive sediment flux. Respective experimental and numerical studies show that it is very difficult to model the flow field correctly if the calibration is only based on field water level measurements. From Fischer and Holleys' (1971) approach, it follows that the lateral dispersive sediment flux from the main channel to the flood plain is reduced in a distorted Froude model and hence, the sedimentation process in the flood plain is underestimated, whereas Rajaratnam and Ahmeds' data (1981) lead to the opposite conclusion. However, the dispersion effect is not so important for irregular cross sections with lateral convective flow at the interface of the main channel and the flood plain.

4. SEPARATING FLOWS

4.1 Steady State Flow (Inland Rivers)

The suspended sediment input into recirculating water zones in a river system (near bank groyne cells, harbours, etc.) exclusively depends on the lateral diffusion, the internal mixing and settling. The internal flow pattern is dominated by the interfacial momentum exchange, the size and the geometry of the water body. There is no local equilibrium of both flow and transport. Therefore, the equilibrium criteria originating from quasi-uniform channel flow cannot be applied. The water exchange coefficient ϵ (Figure 3) shows that the sediment exchange and mixing depends on the water depth to width ratio h/B. Therefore, vertically distorted river models with $X_r < Y_r$ overemphasize the lateral diffusion of sediment and the internal mixing and, presumably lead to an overestimation of the sedimentation.

Similarity problems arise because of the local flow Reynolds number which in a Froude model is much less than in the prototype:

$$Re_m \ll Re_p \sim 10^3 - 10^4 \tag{5}$$

Thus very often a larger scale or a distorted model is required in order to avoid flow relaminarization and to provide the same friction coefficient λ for the recirculating zone.

The Reynolds number of the internal flow is important when detailed information about the sedimentation pattern in the recirculating zone is of interest. For too small a model Reynolds number the mixing cannot be increased by artificial roughness because of the limited flow energy available in the recirculating zone. As the turbulence in the main channel near to the interface has some effect on the lateral diffusion of mass, momentum, and flow energy, the velocity distribution as well as the vertical sediment concentration near to the bank has to be reproduced carefully. For small recirculating zones, such as transverse and longitudinal dykes ($0.1 < h/B < 1$) with negligible bottom friction effect, the lateral sediment transport is dominated by the main flow sediment concentration c. Therefore, the ratio of the actual sediment concentration to the critical value of deposition C_T and C_T', respectively, has to be established in the model.

$$\left(\frac{C}{C_T}\right)_r = 1$$

and (6)

$$\left(\frac{C}{C_T}'\right)_r = 1$$

For extended shallow recirculating zones (river dead arms, etc.), the bottom friction effect is relevant and may have an influence on the configuration of macro-scale gyres and hence, on the sedimentation pattern. If the bottom friction actually influences the internal flow, then additional velocity field data are required for the model calibration.

From the transport capacity concept, it follows that the correct modelling of the sedimentation pattern and time scale requires not just the equality of the Rouse number $(v_s/u_*)_r = 1$ but also the equality of the concentration ratio as expressed by equation (6). The concentration criteria is closely related to the bottom shear relation:

$$\left(\frac{\tau}{\tau_{crit,s}}\right)_r = 1 \tag{7}$$

where $\tau_{crit,s}$ is the critical value for deposition.

In recirculating zones with small exchange interface (groyne structure with small openings or small harbour inlets) and total sedimentation caused by extremely small transport capacity:

$$\left(\frac{C}{C_T}\right) \gg 1 \tag{8}$$

the absolute value of the river concentration is then of secondary importance as it does not affect the sedimentation pattern itself but only the sedimentation time scale according to:

$$\Delta t \cdot \zeta \cdot v_s \cdot c = \Delta Y \tag{9}$$

$$t_r = \frac{X_r^2}{Y_r \cdot c_r \cdot u_r} \quad (\zeta_r = 1) \tag{10}$$

where ζ is the resuspension factor ($p_r = 1$, p the porosity) within the recirculating zone.

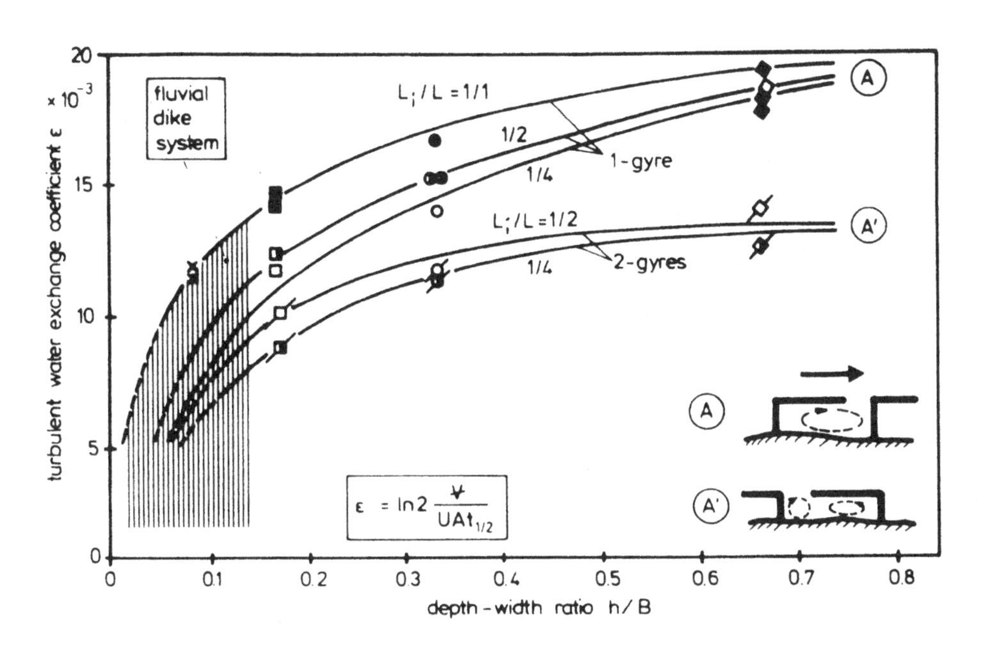

Figure 3. Steady flow mass exchange in confined cells (Westrich, 1977)

4.2 Unsteady Flow (Tidal Rivers)

Recirculating flows in groyne-like structures, harbours, or bays in tidal rivers (or inland rivers with surge waves by navigation traffic) are subject to periodic inflow and outflow causing strong internal mixing. Due to free shear turbulence produced by the lateral convective flow, the internal mixing is much less affected by the water depth to width ratio as shown by some laboratory tracer experiments (Westrich, 1977). Therefore, the sedimentation in tidally driven enclosed water bodies can be satisfactorily modelled in a distorted model, provided the bottom friction has no markable influence on the internal flow and the spiral flow due to streamline curvature is not too dissimilar. However, for a small tidal range ($\Delta h/h$) and shallow water, the bottom flow resistance has to be reproduced correctly. Sedimentation of suspended fine (noncohesive) sediments in these zones can only be handled by physical models if the similarity of the longitudinal and lateral transport of the suspended sediment in the main channel is achieved.

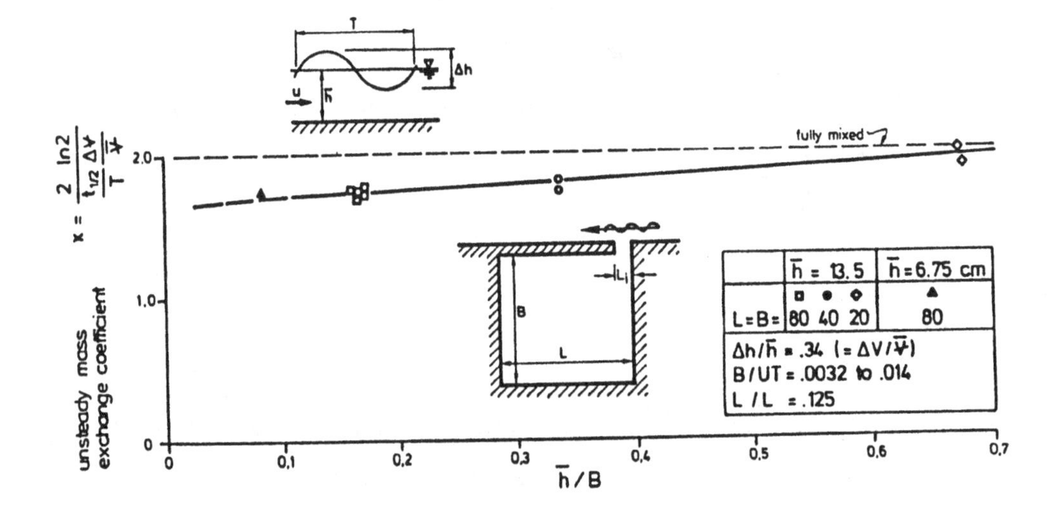

Figure 4. Tidal flow mass exchange in confined recirculating zone (Westrich, 1977).

5. REFERENCES

Fischer, H. B., and Holley, 1971. *Analysis of the Use of Distorted Hydraulic Models for Dispersion Studies.* Water Resources Research.

Itakura, K., August 1980. Open Channel Flow with Suspended Sediments. *American Society of Civil Engineers Hydraulic Division.*

Rajaratnam, N., and Ahmadi, R., 1981. Hydraulics of Channels with Flood-Plains. *J. Inter. Assoc. Hydr. Res.*, Vol 19, No. 1.

Shen, H. W., 1986. *Movable Bed River Models.* IAHR Symp, Scale Effects in Modeling Sediment Transport Phenomena, Toronto.

Yalin, D., 1987. *Determination of a Physical River Model with Mobile Bed.* IAHR Congress, Lausanne.

Westrich, B., 1977. *Water Exchange in Confined Basins Induced by Unsteady Main Stream Currents.* IAHR Congress, Baden-Baden.

Westrich, B., and Juraschek, M., 1985. *Flow Transport Capacity for Suspended Sediment.* IAHR Congress, Melbourne.

Westrich, B., 1988. *Fluvial Sediment Transport - Effects on the Morphology and Significance for Water Quality.* Oldenbourg, Munchen, Berlin (in German).

Wiuff, R., May 1985. Transport of Suspended Material in Open and Submerged Streams. *American Society of Civil Engineers Hydraulic Division.*

LIMITS AND POSSIBILITIES WITH ARTIFICIAL MODEL BED LOAD MATERIAL

PETER LARSEN
Institut für Wasserbau und Kulturtechnik
Universität Karlsruhe
Kaiserstrasse 12
D-7500 Karlsruhe 1
Federal Republic of Germany

1. INTRODUCTION

Scale selection when modeling sediment transport using artificial material is discussed. Model studies in the Theodor Rehbock Laboratory some years ago required modeling of a mountain river carrying heavy sediment loads. To accomplish this in a comparatively small-scale model a plastic material was used. To make possible an interpretation and extrapolation of test results, comprehensive studies of the hydraulic and sedimentologic properties of the material were undertaken. Based on the results of these studies and on theoretical considerations, possibilities and limitations of scale selection were derived.

2. THEORY

2.1. Sediment Properties

Basic to the following considerations is the assumption that grain diameter is significant in two ways: one, as a measure of hydraulic roughness and another as characteristic of sediment transport. For a given material, either uniform
or graded, the two significant diameters have, in general, different values.

This fact can be used in combination with differences in density between model and prototype material to choose the model scale ratio to advantage.

With uniform grains, the equivalent roughness, de, is the diameter of uniform, rounded sandgrains, as used by Nikuradse, which give the same friction loss (or shear stress, or velocity distribution) as obtained with the grains in question. With a graded material a certain grain size is assumed to be representative of the equivalent sand roughness. In this presentation the suggestion of Strickler is adopted:

H. W. Shen (ed.), Movable Bed Physical Models, 107–114.

$$n = d_{90}^{1/6}/26 \tag{1}$$

with n = Manning coefficient
d_{90} = fraction for which 90% is finer.

With uniform material, the sedimentation diameter is the mean diameter of the grains. Following Meyer–Peter's and Muller's approach, the significant diameter of a graded material is:

$$d_m = \sum \frac{p_i d_i}{100} \tag{2}$$

with d_i the average diameter of grains in the range d_{i-} to d_{i+} and p_i the percentage by weight of the fraction in this range, the summation covering all sizes.

2.2. Hydraulic Similitude

For an undistorted model, the energy slope is required to be the same as in the prototype. Energy loss has two contributions: (1) loss due to skin friction, and (2) loss due to bed relief.

1) Hydraulic similarity requires relative roughness to be the same in model and prototype. With uniform flow, this results in similar velocity distribution and thus similar bed shear stress. This, however, is an approximation when a bed relief is present.

2) Fredsöe (1982) relates bed form to the grain Froude number:

$$F_* = \frac{\tau}{g(\rho_s - \rho)d_m} = \left(\frac{u_*^2}{g\frac{\rho_s - \rho}{\rho}d_m} \right) \tag{3}$$

with dm as representative geometric length and τ the shear stress due to skin friction. He shows that for $F_* < 0.225$ bed form is a unique function of F_*.

Assuming that the energy loss due to bed form depends only on geometric similitude (sufficiently high Reynolds number), a sufficient requirement thus is:

$$(F_*) = 1 = \frac{(\tau)_r}{(\rho_s - \rho)_r (d_m)_r} \tag{4}$$

Since $(\tau)_r = (L)_r$, it follows that:

$$(\rho_s-\rho)_r\ (d_m)_r = (L)_r \tag{5}$$

This condition together with the one for skin friction:

$$(d_{90})_r - L_r \tag{6}$$

then constitute the hydraulic similitude requirements.

2.3. Sediment Transport

The Meyer-Peter and Müller equation is widely accepted. It was the result of extensive measurements over a 16-year time period and encompassed slopes in the range from 4.10-4 to 2.10-2, densities in the range from 1250 to 4000 kg/m3, and effective grain diameters between 0.4 and 30mm.

In dimensionless form the Meyer-Peter and Müller equation can be written:

$$g_s* = 8F_* \left[\left(\frac{n}{n_r}\right)^{1.5} - \frac{0.047}{F_*}\right]^{1.5} \tag{7}$$

with g_s^* = unit width dimensionless sediment transport rate; F_* = grain Froude number

$$\frac{\rho \cdot u_*^2}{(\rho_s - \rho) g d_m}$$

n_r = Manning coefficient for the river bed; n = Manning coefficient for skin friction, Equation 1.

The factor n/nr takes account of the fact that only the skin friction contributes to bed load transport.

2.4. Approach To Model Scaling

Eisenhauer (1986) makes the following propositions:

1) Skin friction is governed by d_{90} of the bed material;

2) Sediment transport is governed by a representative grain diameter d_m as overall characteristic;

3) Meyer-Peter and Müller equation contains the hydraulic and sedimentologic importance of d_{90} and d_m.

These assumptions are equivalent to the above derived equations 5 and 6. From these equations follows that with an undistorted model and using sediments of the same density as in the prototype, two points on the sieve curves of model and prototype must be proportional. The definition of the d_m-points suggest that the two sieve curves will be similar with the ratio $(L)_r$ between grain diameters of equal percentage.

Using a lighter material in the undistorted model requires a dm which is correspondingly less reduced according to equation 5. Since d90 and dm are differently scaled, the model sieve curve becomes a distorted version of the prototype sieve curve. Eisenhauer defines a non-uniformity coefficient:

$$U = d_{90}/d_m \tag{8}$$

which, when applied to the distortion of the sieve curve required by equation 5, yields a distortion factor:

$$n_s = \left(\frac{d_{90}}{d_m}\right)_r = (\rho_s - \rho)_r \tag{9}$$

Figure 1 shows as an example a prototype sieve curve and two distorted sieve curves.

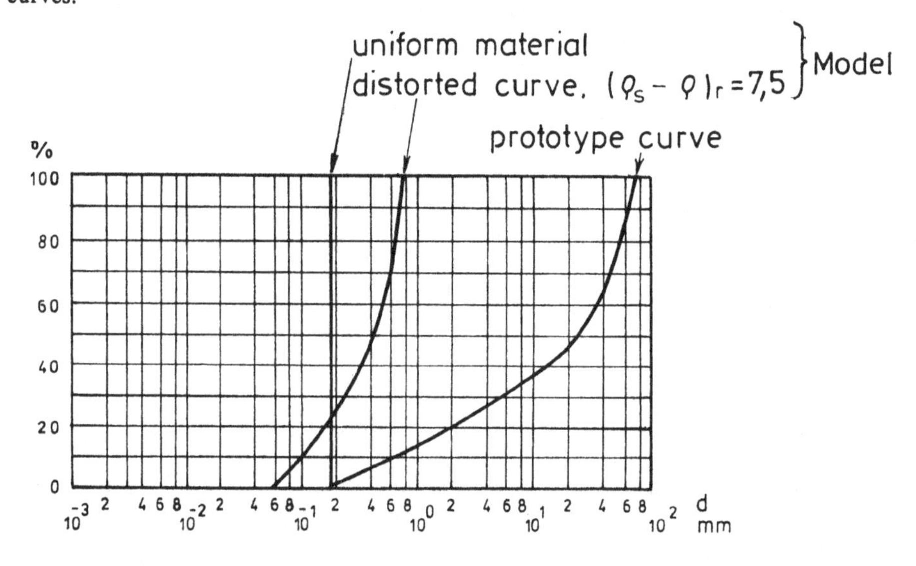

Figure 1. Example of sieve curves.

2.5. Limiting Conditions

The Meyer-Peter and Müller equation does not take account of viscosity effects in the boundary layer at the bed. This means that the grain Reynolds number, R_*, should be above

70. Adopting d_{90} as significant grain diameter, the limiting condition for the model becomes:

$$R_* = \frac{u_* \cdot d_{90}}{\upsilon} > 70 \tag{10}$$

This ensures that the turbulence be fully developed and thus the friction losses and the initiation of sediment transport are independent of viscosity.
Since $u_* = U\sqrt{f/8}$ where U is the depth averaged velocity, and using the definition of Re, this requirement is equal to:

$$\frac{\sqrt{f} \cdot Re d_{90}}{4y} > 200 \tag{11}$$

with f = Darcy-Weisback friction factor; Re = 4U . y/υ; U = depth averaged velocity; y = depth; υ = kinematic viscosity.

Equation 11 states the limiting condition in the Moody diagram for fully developed turbulence, i.e., Reynold number independency.

When model flow rate is scaled according to the Froude law, equation 11 yields an upper bound of the model scale ratio.

$$(L)_r \leq \left(\frac{\sqrt{f}\, Re d_{90}/4y}{200}\right)^{2/3} \tag{12}$$

with prototype values on the right hand side.

Equation 10 is also equivalent to the following statement regarding the dimensionless grain diameter of the model:

$$D_* > 45 \tag{13}$$

with $D_* = (R_*^2/F_*)^{1/3}$ = dimensionless grain diameter according to Bonnefille (1963). Equation 13 is claimed to separate Reynolds dependent from Reynolds independent regions in the Shields' diagram.

From Equation 13 a maximum scale ratio for the effective grain diameter, d_m, is obtained:

$$(d_m)_r \leq \frac{D_*}{45(\rho_s - \rho)_r^{1/3}}$$

where D_* is the dimensionless grain diameter of the prototype.

The distortion of the sieve curve also presents a limiting factor to scale selection: The distortion cannot exceed that which is obtained with uniform grains in the model.

For practical use the stated conditions are shown graphically in Figure 2. With the aid of this figure it is possible to select model ratios (L)r and $(\rho_s - \rho)_r$ when prototype conditions are known.

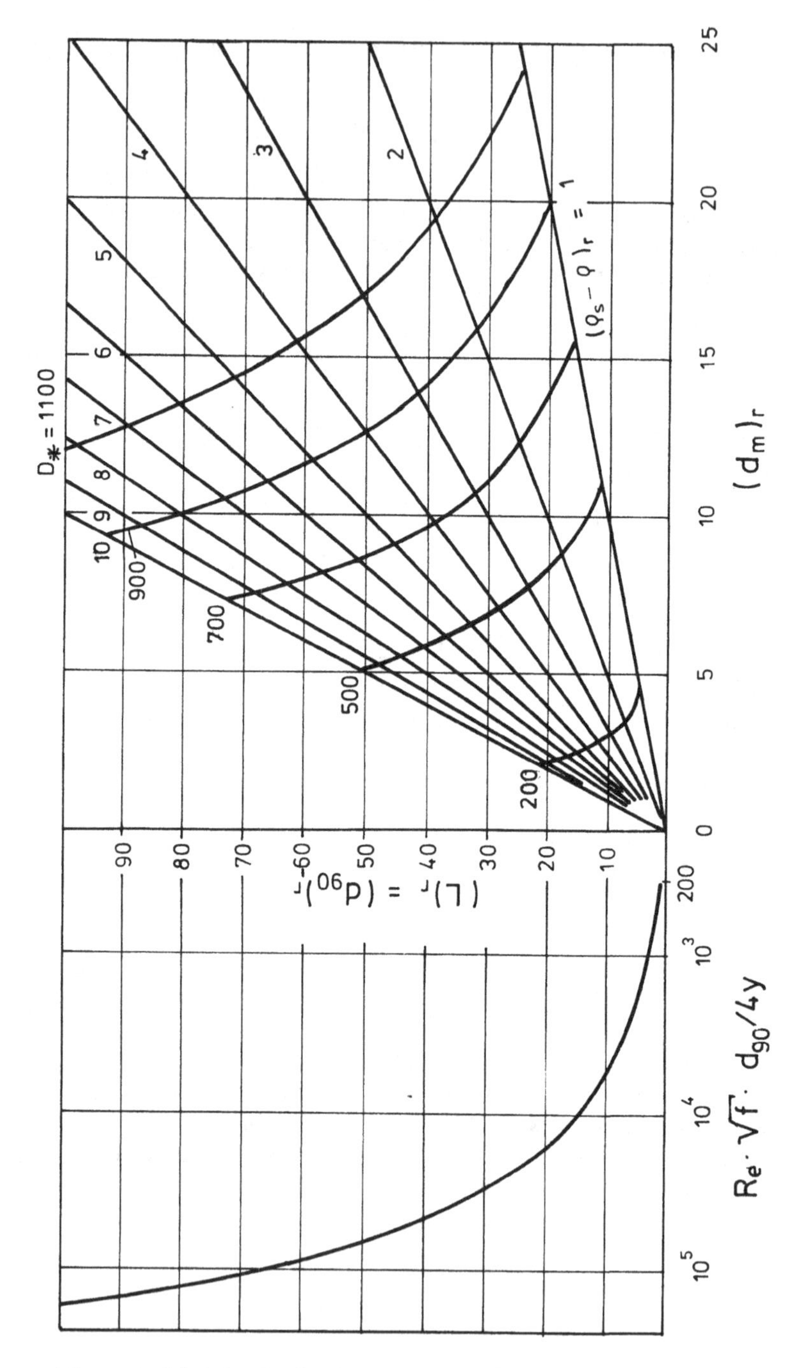

Figure 2. Model ratios for length and density scales.

2.6. Summary of Results

A method developed at the Theodor Rehbock Laboratory, University of Karlsruhe, is described. The selection of model scale ratios was satisfactory when using the method in modeling a gravel bed river. Two graphs are presented which enable scale selection based on information on prototype conditions.

3. SYMBOLS AND DEFINITIONS

d_e = roughness height

d_{90} = 90% by weight is finer

d_i = grain diameter

d_m = significant grain diameter in Meyer Peter and Müller transport equation

D_* = dimensionless grain diameter

F_* = grain Froude number

f = Darcy–Weisbach friction factor

g_s^* = dimensionless sediment transport rate

n = Manning coefficient

n_s = distortion factor for sediment

p_i = percentage finer than i

R_* = grain Reynolds' number

U = depth averaged velocity

u_* = shear velocity

y = depth

ρ = density

υ = kinematic velocity

$(\ )_r$ = denotes ratio of prototype to model

4. REFERENCES

Eisenhauer, N. O. 1986. Alternative Concept for Scaling Hydraulic Models with Movable Bed. Proceedings, IAHR Symposium on Scale Effects, Toronto, Canada.

Fredsöe, J. 1982. Shape and Dimensions of Stationary Dunes in Rivers. Journal of Hydraulic Division, ASCE, 108.

Zanke, U. 1982. Grundlagen der Sedimentbewegung. Springer-Verlag Berlin-Heidelberg-New York.

SURVEY OF LIGHTWEIGHT SEDIMENTS FOR USE IN MOBILE-BED PHYSICAL MODELS

ROGER BETTESS
River Engineering Department
Hydraulics Research
Wallingford
Oxfordshire OX10 8BA
United Kingdom

ABSTRACT. This paper describes a survey carried out to collect information on lightweight sediments of use in mobile-bed modelling. The materials are listed and the specific gravity and size range for each material is provided. Comments on the materials by various laboratories are included.

1. INTRODUCTION

As part of the work of the IAHR Task Force on the scaling of mobile-bed models, it was decided to collect information on the use of and experience with lightweight sediments in mobile-bed physical models. It was decided that this could be most usefully achieved by circulating a questionnaire to as many workers in the field as possible asking for information. Hydraulics Research agreed to be responsible for this work.

2. QUESTIONNAIRE

A questionnaire was formulated with the advice of Dr. Klaassen of Delft Hydraulics and was sent out in May 1987. A copy of the questionnaire accompanies this paper.

3. RESPONSE TO QUESTIONNAIRES

Twenty-seven completed questionnaires have been returned. Altogether sixteen institutions in eleven different countries replied. The geographical spread of the response was wide though the largest number of replies was from European countries (6).

The information has been arranged according to material types. The values quoted for specific gravity and sediment sizes are derived from the questionnaires and the author disclaims any responsibility for values that are incorrect or misleading. Similarly

H. W. Shen (ed.), Movable Bed Physical Models, 115–123.

the comments on the materials are derived from the questionnaires and their inclusion does not indicate agreement or disagreement by the author. Where possible, an indication is given of the size of material that has been used in investigation. In some cases a single sediment size was given in the response to the questionnaire in which case this is given with no further qualifications. In some cases it was indicated that this was the D_{50} size, in which case this is specified. For those cases in which a grading curve was provided, the D_{10} and D_{90} sizes are given. Where provided, an estimate of the practical range of sizes is given. The information on size and specific gravity is summarized in Figure 1. Any material points representing size and specific gravity used in an investigation were joined by a straight line. For materials with a fixed specific gravity, a single line results; for materials with a range of possible specific gravities, an area is enclosed. If the possible range of sizes extends further, it is indicated by a dashed line. Table 1 provides a summary of the materials available for different ranges of specific gravities.

For porous materials the effective specific gravity is for the water saturated grains. In some cases the difference between the 'saturated surface dry' specific gravity (ssd) and the fully dried specific gravity can be very large, for example:

	Saturated Surface Dry Specific Gravity	Dry Specific Gravity	Comment
Bakelite	1.38	1.49	due to saturation of filler
Lytag	1.71	1.92	porous solid

(P. R. Kiff, private communication)

3.1. Acrylonitrile Butadienne Styrene (Ter Polymer) (ABS)

Replies: 1
Specific gravity: 1.22
Sediment size used: 2.3 mm to 3.0 mm
Comments:

1. Problems of adhering air bubbles was solved by adding a detergent.

3.2. Bakelite

Replies: 3
Specific gravity: 1.3 to 1.45
Sediment size used: $D_{10} = 0.3$ mm
$D_{90} = 4.0$ mm

Comments:

1. Seems no longer to be manufactured, laboratories rely upon existing stocks but no new sources of supply known.
2. Shows good resistance to wear when being pumped.
3. Tends to rot in water.
4. Noticeable reduction in diameter over a long period due to weathering.
5. Can be difficult to wet. If used in models where varying water levels leads to some of the material drying out, dry material has a tendency to float.

3.3. Bakelite and Gravel

Replies: 1
Specific gravity: Bakelite = 1.35
Gravel = 2.65
Sediment size used: Bakelite, $D_{50} = 0.85$ mm
Gravel, $D_{50} = 4$ mm

Comments:

1. Intention was to simulate the suspended load by the Bakelite and the bed load by the gravel. The sediment was made up of 60% gravel and 40% bakelite.

3.4. Coal

Replies: 5
Specific gravity: 1.37 to 1.61
(note: 1.6 is a high density for coal, more usual range 1.35 to 1.45)
Sediment sizes used: $D_{10} = 0.33$
$D_{90} = 4$ mm
Practical range of sizes: 0.1 to 40 mm

Comments:

1. Larger sizes can tend to have a "flaky" particle shape. This may lead to hydraulic segregation of different particle shapes in model though no size variation observed.
2. Slight inhomogeneity of specific gravity.
3. Absorption of algae may take place in open-air models.

3.5. Lightweight Aggregate

Replies: 1
Specific gravity: 1.7
Sediment sizes used: 1 to 3 mm
Comments:

1. The material is sintered fly-ash and is marketed under the name LYTAG. A range of sizes and specific gravities is obtainable.

3.6. Perspex

Replies: 2
Specific gravity: 1.18 to 1.19
Sediment sizes used: D_{50} = 0.3 to 0.37 mm
Practical range of sizes: up to 1 mm
Comments:

1. Becomes very dusty when swept from model surface.

3.7. Polyamidic Resins (Nylon)

Replies: 1
Specific gravity: 1.16
Sediment sizes used: 0.2 mm to 4 mm
Practical range of sizes: 0.1 mm to 5 mm
Comments:

1. Shape of coarser grains is different from fine gravel.

3.8. Polystyrene

Replies: 5
Specific gravity: 1.035 to 1.05
Sediment sizes used: D_{50} = 1 to 2 mm
D_{90} = 3.1
Practical range of sizes: 0.5 mm to 3.2 mm
Comments:

1. Difficult to wet; dirt or fat on the grains can cause them to float though washing with soap may prevent this.
2. Cohesion of moist material - a special sediment feeder was developed.
3. No cohesion.
4. Durability good.
5. No reduction of diameter.
6. No pollution.
7. Weathering problems could be avoided by using additives.

3.9. PVc

Replies: 3
Specific gravity: 1.14 to 1.25
Sediment sizes used: 1.5 to 4 mm
Comments:

1. Hydrophobic.
2. A wide range of specific gravities can be obtained by altering the volume and density of the filling material though the cost is normally prohibitive.

3.10. Sawdust treated with Asphalt

Replies: 1
Specific gravity: 1.05
Sediment size used: 0.6 to 1.0 mm
Comments:

1. Flocculation occurred during experiment.
2. If stored in sunlight the material stuck together.

3.11. Sand of Loire

Replies: 1
Specific gravity: 1.5
Sediment size used: 0.96 mm
Practical range of sizes: 0.63 to 2.25 mm
Comments:

1. Presence of dust.

3.12. Silica, Ground

Replies: 1
Specific gravity: not reported
Sediment size: not reported
Comments:

1. Color sometimes does not provide enough contrast for filming.

3.13. Walnut Shells, Ground

Replies: 1
Specific gravity: 1.33
Sediment size used: 0.27 mm
Practical range of sediment sizes: 0.15 to 0.41
Comments:

1. Deteriorates in 2 to 3 months.
2. Tends to change water color to dark brown.

3.14. Wood, Granulated Obeche Wood

Replies: 1
Specific gravity: 1.10
Sediment size: 0.8 mm
Comments:
1. Difficult to prepare.
2. Deteriorates in water.

4. POSSIBLE MATERIALS

The following materials have been considered for use in mobile-bed models but, as far as the author is aware, never been used.

	Specific Gravity	Sediment Size
Rigid PVc	1.37	D_{10} = 0.24 mm D_{90} = 0.84 mm
Araldite resin	1.14	D_{10} = 0.24 mm D_{90} = 0.41 mm
Virgin Diakon Polymer	1.20	D_{10} = 0.06 mm D_{90} = 0.10 mm
Sorbitex	dry 2.11 ssd 1.49	D_{50} = 1.5 mm
Lec	dry 1.46 ssd 1.36	D_{10} = 1.1 mm D_{90} = 2.5 mm

(P. R. Kiff, private communication)

5. GENERAL COMMENTS ON USE OF LIGHTWEIGHT MATERIALS

A number of replies commented on the use of lightweight materials in general.

1. Lightweight sediments are inapplicable where sediment extends above water surface, for example in coastal models, for breaking waves and where sediment particle accelerations are significant.
2. Lightweight sediments are only needed for problems of bed-load and initiation of motion.

6. FURTHER INFORMATION

The author is prepared, at least in the short-term, to act as a clearing house for information on lightweight sediments. He would like to invite anybody who accidentally was not included in the original survey to send him the appropriate information. If warranted, an updated register of lightweight sediments might be produced in the future.

7. CONCLUSIONS

A survey of the use of lightweight sediments in the mobile-bed models has been carried out. A total of twenty-seven questionnaires described fourteen different materials that have been used in mobile bed model investigations. The information has been summarized accordingly to material and range of specific gravity.

8. ACKNOWLEDGEMENTS

The author would like to thank all those who responded to the questionnaire. He would also like to thank Mr. P. R. Kiff of the Sedimentation Laboratory of Hydraulics Research for valuable help and advice.

Table 1

Specific Gravity	Material
1.0 to 1.1	Polystyrene, sawdust with asphalt
1.1 to 1.2	Nylon, Perspex, PVc, Wood
1.2 to 1.3	ABS, Bakelite, PVc
1.3 to 1.4	Bakelite, Coal, Walnut shells
1.4 to 1.5	Bakelite, Coal, Sand of Loire
1.5 to 1.6	Coal, Sand of Loire
1.6 to 1.7	Lightweight aggregate

IAHR Task Force on Scaling of Mobile-bed Models
Questionnaire on Lightweight Materials for Use in Mobile-bed Models

1. Laboratory ____________________
2. Material ____________________
3. Density:
   Dry density ____ Bulk density ____ Saturated-surface dry density ____
   (see definitions overleaf)

4. Size:
   Size used in investigation ____________________
   (grading curve if available)

   Range of practicable sizes: Estimated largest size ____________________
   Estimated smallest size ____________________

5. Supply:
   Source of supply (where appropriate) ____________________
   Cost ________________

   Type of investigation for which material has been used ____________________

6. Practical Problems Encountered:
   (For example, durability, reduction of diameter in time due to wear, cohesion, pollution, weathering, measurement problems)

7. Any Other Comments:

* * * * * *

1. Dry density = specific gravity
   = weight of solid/volume of solid (including closed pores)
2. Bulk density = weight of solid/volume of packed bed in air
3. Saturated surface dry density = weight of solid + weight of water within solid/volume of solid + water within solid
   (only applicable to porous solids)

Please return to:

Dr. R. Bettess
Hydraulics Research
Wallingford
Oxfordshire
OX10 8BA
United Kingdom

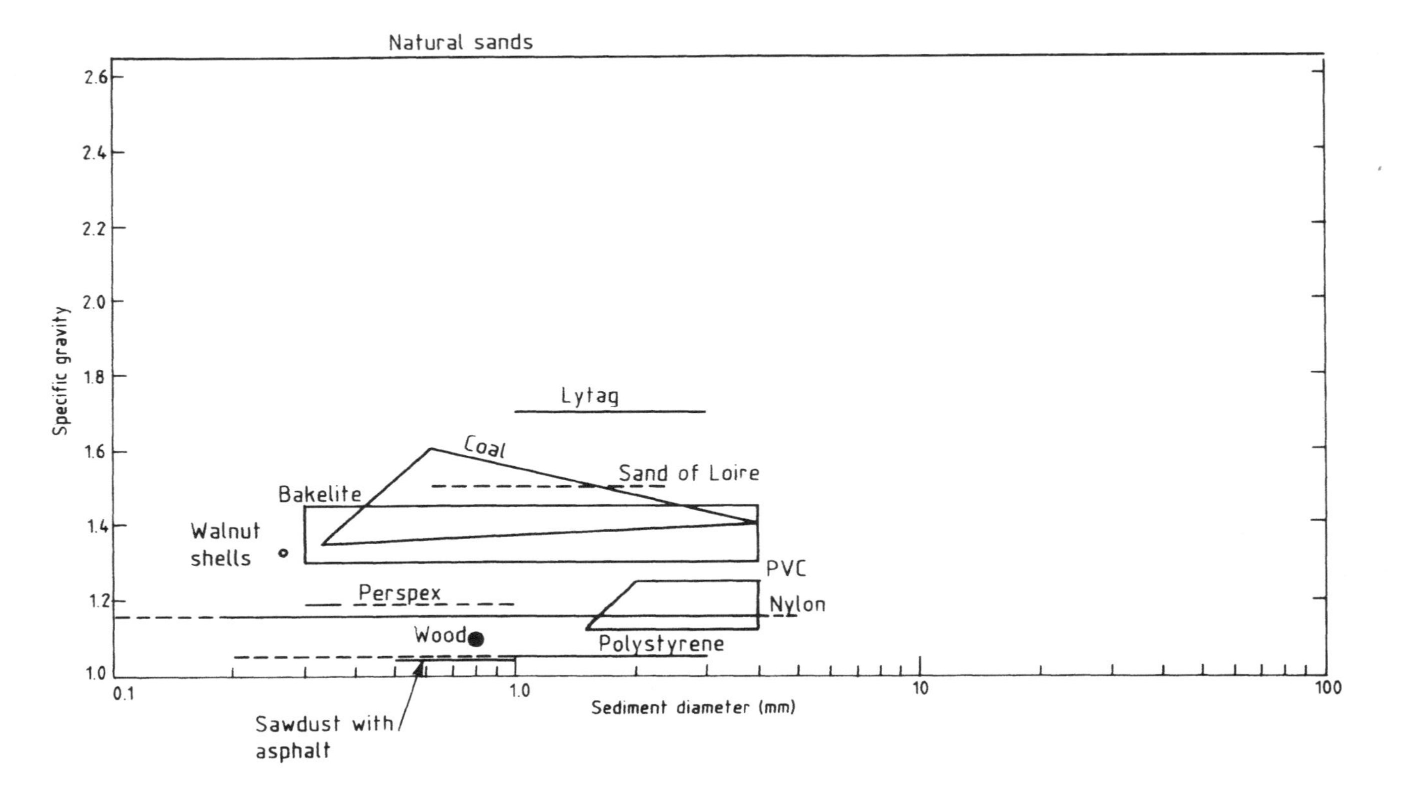
Natural sands
Specific gravity
2.6
2.4
2.2
2.0
1.8
1.6
1.4
1.2
1.0
Lytag
Coal
Sand of Loire
Bakelite
Walnut
shells
PVC
Perspex
Nylon
Wood
Polystyrene
0.1
1.0
10
100
Sediment diameter (mm)
Sawdust with
asphalt

EXAMPLES FOR USING SAND AND LIGHTWEIGHT MATERIAL IN MOVABLE BED MODELS

HANS-J. VOLLMERS
Universität der Bundeswehr München
D-8014 Neubiberg
Federal Republic of Germany

ABSTRACT. Up to now the simulation of sediment transport processes in suitable models is difficult. The advantage of physical models is that the interaction between liquid and sediment can be directly observed. In this paper the basic differences are demonstrated if one uses sand or lightweight granulates in physical models. The second case demands different scales in horizontal and vertical dimensions. The similarity laws for a rough determination of the model scales are presented. The most important installations for running sand and lightweight material models are discussed.

1. INTRODUCTION

Up to now, it is still difficult to explain sediment transport processes physically. According to that, it's not possible making up an exact forecast in using mathematical or physical models.

Physical models have the advantage of reproducing the altering effects between water and sediment. So obviously the problem seems to consist of developing similarity laws and good model techniques.

This problem is well-known for quite a long time, and we should ask why we still conduct a workshop on models with movable bed. I will try to explain this shortly:

Eighty-five years ago hydraulic modelling had been started. Since this time physical models after Froude and Reynolds were used. Concerning the questions of sediment transport, we improvised, following the motto: "Necessity is the mother of invention," e.g. with sand or lignite etc. These testing results had only a qualitative character. You can change the well-known proverb, "Necessity knows no law" in "Necessity knows no similarity law!". Anybody knows this, but nobody says it or publishes it!

Since Shields, very much research had been done to define similarity laws, which would allow to interpret the model results also quantitatively. Nevertheless, most of the hydraulic laboratories develop their own similarity laws and their own model technique to solve these practical problems. During a visit to a Chinese hydraulic laboratory in 1983 (on the occasion of the "2nd International Symposium on River Sedimentation" in Nanjing) we were shown a big movable-bed-model of the Yangtse-River, but the question concerning the similarity law had not been answered (see above "Necessity knows no similarity law").

Unfortunately, nobody says or publishes anything about these "crutches" which we use in hydraulic modelling, especially the model technique. It should be mentioned that the best similarity law isn't worth anything if there is no possibility to prove it on a real model.

H. W. Shen (ed.), Movable Bed Physical Models, 125–140.

2. SIMILARITY CONSIDERATIONS

The well-known diagram from Nikuradse, showing the relationship between the constant C and the Re^* number, is certainly still the basis of most of the similarity criteria. If the Re^* number is greater than 70, the sediment transport depends only on the Froude-number and the relative roughness h/d. The Froude-number and h/d differ only in the factor I.

$$Fr'^* = \frac{v_o^{*2}}{g \cdot d} = \frac{h}{d} I \tag{1}$$

Günzel (1964) demonstrated this relationship for a practical case by adding the density factor ρ' $(\rho_s - \rho_w)/\rho_w$. He then got the similarity relation

$$\frac{\rho'_m}{\rho'_n} \frac{h_n}{h_m} I_n = \frac{d_n}{d_m} I_m \tag{2}$$

He concluded from this that $I_n \neq I_m$, if $\rho'n/\rho'_m \neq 1$. There is no geometrical similarity if you change the density of the model material (using water as model fluid!). In other words, you have to distort the model if you do not use sand as model material.

Finally, Günzel set up the following similarity condition for a possible geometrical reduction of the natural sediment

$$\frac{\sqrt{g \cdot h \cdot K \cdot I}\; d_{50n} \cdot K}{\nu} \geq 70 \tag{3}$$

h water depth for initiation of sediment motion [m]
I slope [l]
d grain diameter in the nature [m]
g gravity acceleration $[ms^{-2}]$
ν kinematic viscosity $[m^2s^{-1}]$
K d_{50m}/d_{50n} [1]

For a practical case Günzel found out with
h = 1m, d = 0.02m, I = 0.005, $\nu = 10^{-6}\ m^2/s$ (20°C)

$$\frac{\sqrt{9.81 \cdot 1 \cdot K \cdot 0.005}\; \cdot 0.02 \cdot K}{10^{-6}} \geq 70 \tag{4}$$

and $63.3 \cdot K^{3/2} \geq 1 \rightarrow K \geq 1 : 15.9$

For the practical realization of the model you still need many considerations, e.g., surface available, etc. Choosing the scale, one has a certain variability by changing the water temperature. In this context let us remember the heating effect avoiding ripples in the model (sand with d ≈ 0.6 mm).

In most of the practical cases a geometrical reduction of the bed material won't be possible. Following the Shields diagram, the dimensionless parameters are situated in a section where a relation between Re^* and h/d still exists. Especially in tidal rivers one can find a grain size distribution with a characteristic diameter of 0.2 to 0.6 mm. A geometrical reduction is excluded because the very high scale numbers would demand a silt-clay material in the model. Such material is fully cohesive. On the other hand such models have to be distorted (surface available) and thus light-weight material will be required.

For the above mentioned section in the Shields diagram, Gehrig (1967) developed his well-known similarity criteria which you can find in Figures 1 and 2. However the main ideas of Günzel and Gehrig correspond a lot.

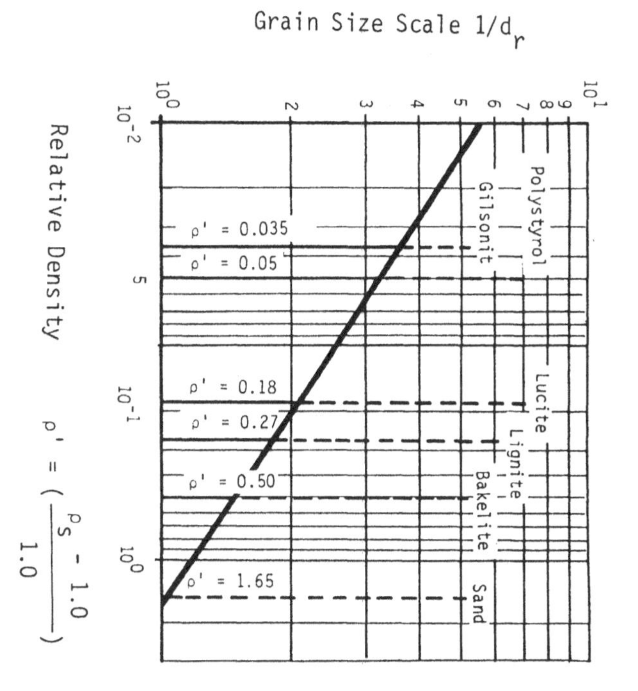

Figure 1. Scale relations for movable bed models with consideration of the roughness condition.

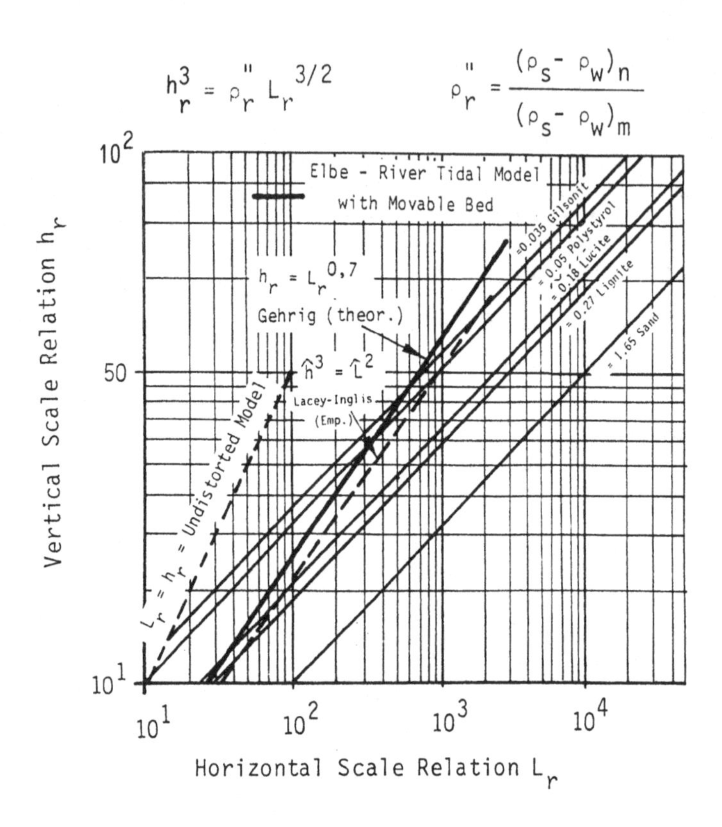

Figure 2. Scale relations between density and grain diameter of the model bed material.

3. CASE STUDIES

3.1. Sand Models

3.1.1. *Neumagen (South Black Forest).* In constructing a sand model with geometrical reduction of the natural bottom material, there will be for the most part bed load transport. If you furthermore assume that this should be a new river training you can hardly use sediment transport measurements. You then have to follow hydraulical sedimentological calculations. In such cases the use of the Meyer–Peter formula in consideration of the limits (e.g. grain size distribution) will be recommended.

Figures 3 and 4 contain the main data of the river. One expected sedimentation downstream from the break of the slope. Therefore it was proposed to construct a sediment trap in this place of the river (Figure 4).

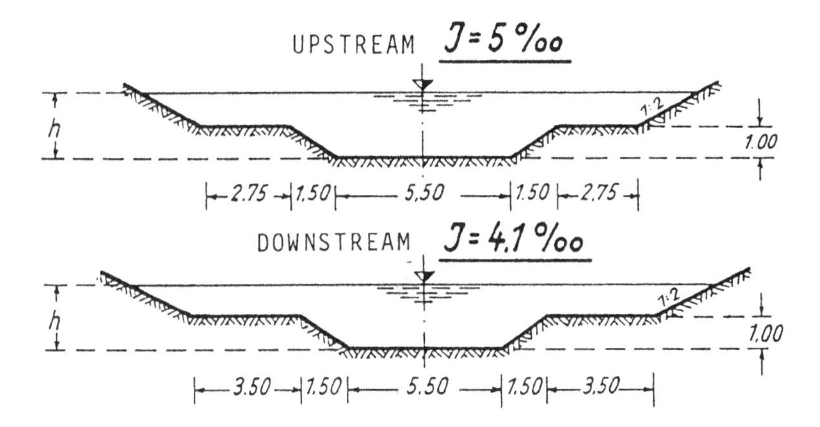

Figure 3. Cross section upstream and downstream from the sand trap.

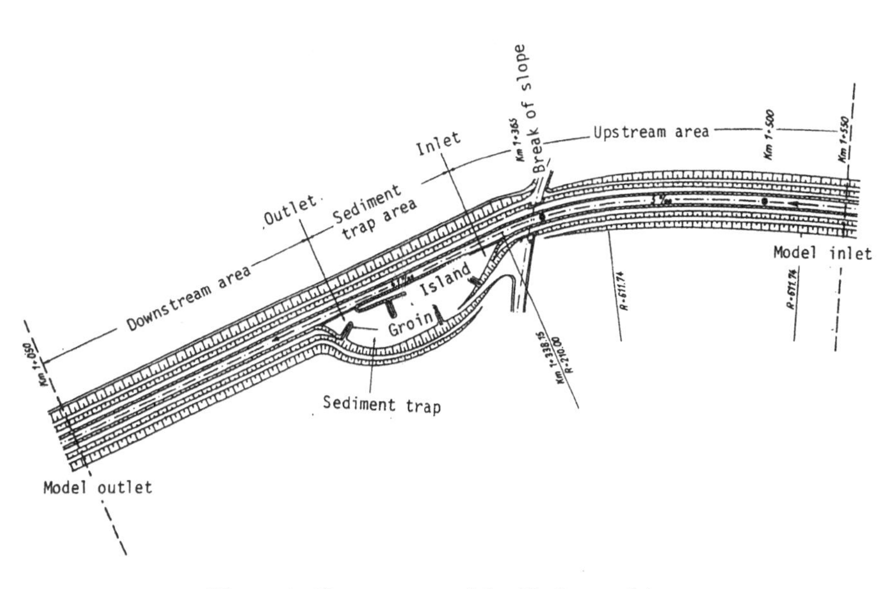

Figure 4. Neumagen model with the sand trap.

Figure 5 shows the grain line of the bottom material from the Neumagen in comparison to the basic grain curves of Meyer–Peter's formula.

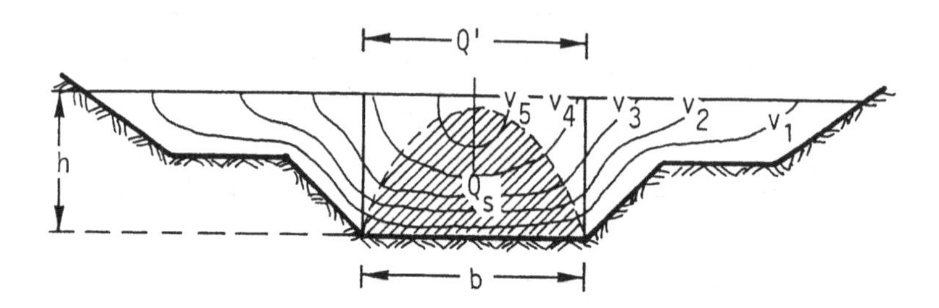

Figure 5. Grain distribution curve of the Neumagen River in comparison to Meyer–Peter.

In regard of the correspondence, an application of the formula can be accepted. It was applied in the following form

$$m_G = \frac{8}{g}\sqrt{\frac{1}{\rho_w}}\ (\rho_w\, g\, I_r\, R_s - 0.047(\rho_s - \rho_w)\, g\, d_m)^{3/2} \left[\frac{kg}{m\ \ s}\right] \quad (5)$$

I_r friction slope [1]; $I_r = I(k_{St}/k_r)^{3/2}$
k_{St} roughness coefficient after Stricler $[m^{1/3}/s]$
k_r coefficient for the grain roughness $[m^{1/3}/s]$; $k_r = 26/(d_{90})^{1/6}$
R_s hydraulic radius [m], related to the cross-section (responsible for the bed load transport); $R_s = h \cdot Q_s/Q'$
Q_s discharge, responsible for the bed load transport $[m^3/s]$
Q' discharge above the bottom $[m^3/s]$ (Figure 6).

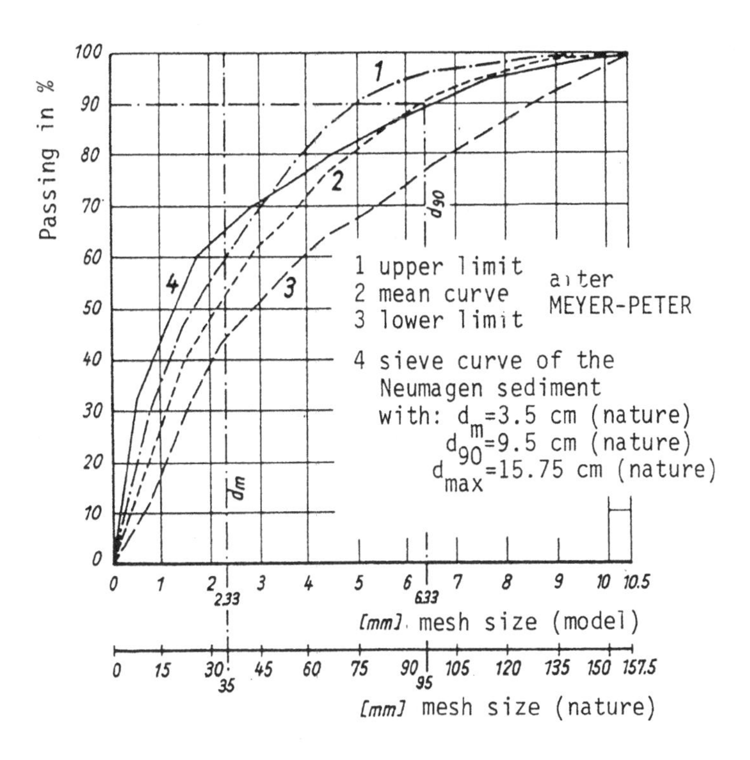

Figure 6. Q_s (shaded area).

The relation Q_s/Q had been calculated from the isolines of the velocities measured in the model. In Figure 7 you can see that the simplified assumptions $Q_s/Q = 1$ and $= b/(b+2h)$ do not give sufficient results.

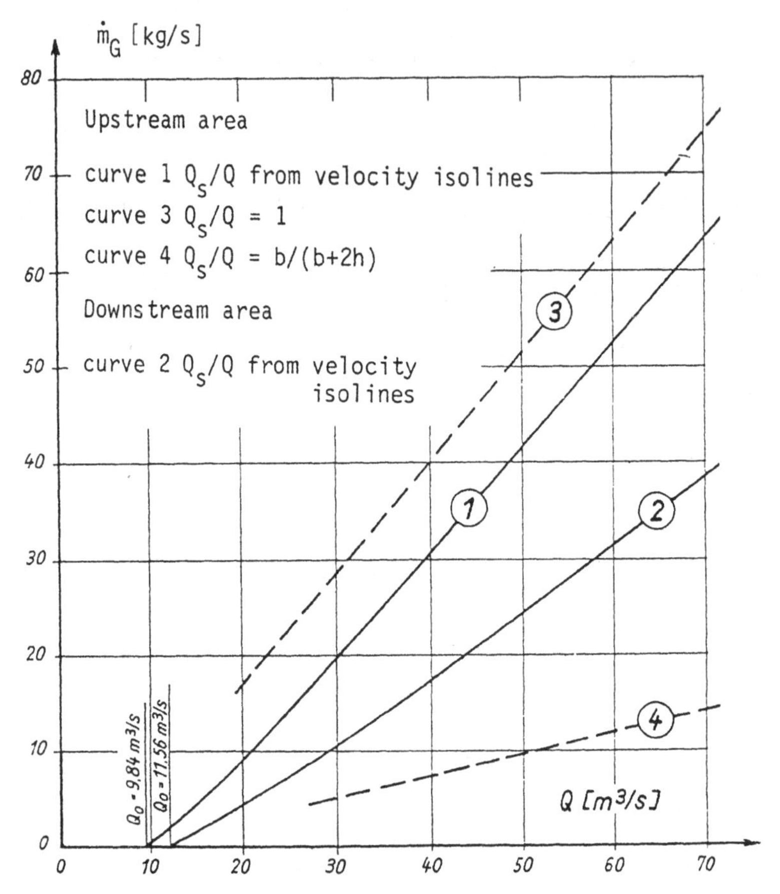

Figure 7. Bed load transport curves in the Neumagen River.

The model was constructed undistorted of 1 : 15. The model material was composed of 10 different fractions geometrically reduced and had been mixed up carefully. The bed load material supply took place by means of the installation shown in Figure 8. By changing the velocity of the conveying belt the sediment volume could be adapted to the discharge.

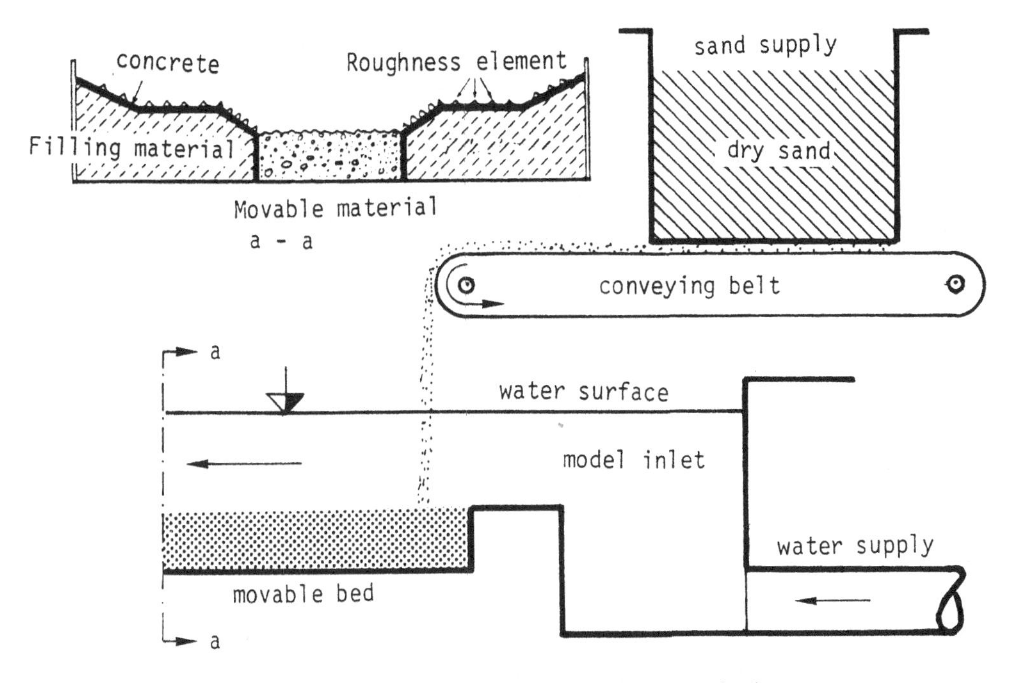

Figure 8. Water and sand supply of the Neumagen model (schematical).

Because the main bed load transport starts during higher discharges, different hydrographs had been investigated in the model.

In Figure 9 you can see the grain size distribution collected by the sand trap at the downstream bottom and the upstream bottom (original bed load). The separation effect is clear. The fine material is stored in the trap, the downstream distribution shows the coarser material. To analyze the sieve curves one used masses between 220 and 950 kg.

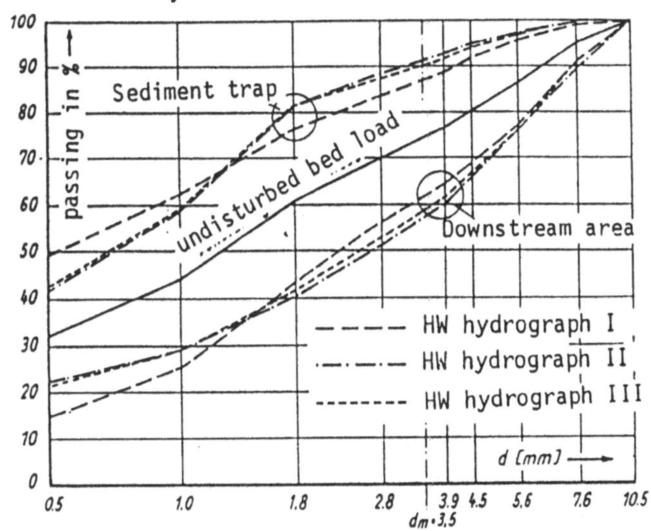

Figure 9. Grain distribution curves of the original and the separated bed load.

The pre-calculated bed load transport could be verified concerning the measured masses.
The model investigations were performed in the "Theodor-Rehbock-Flußbaulaboratorium" at the University of Karlsruhe in 1961–62.

3.1.2. *Weissach (Alpes).* Today, the Laboratory for Hydromechanics and Hydraulic Structures of the University of the Federal German Armed Forces Munich is performing the first tests in a sand model with geometrical reduction of the natural sand material. The model scale is 1 : 20. The situation looks a bit like the one of these in the Neumagen River (Figure 10).

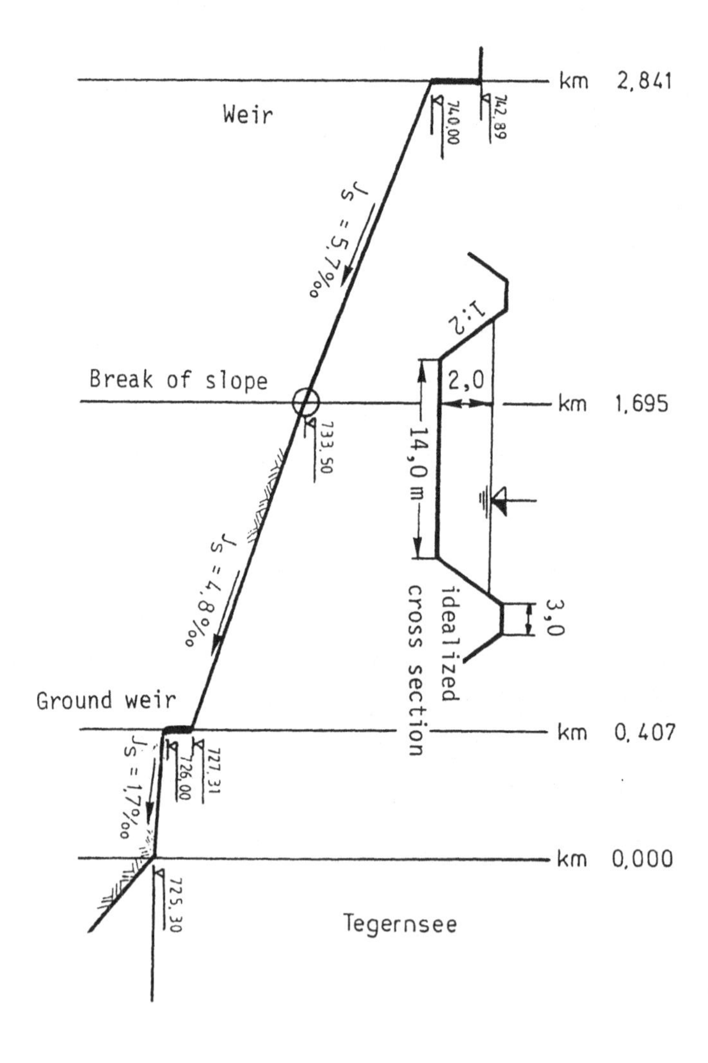

Figure 10. Longitudinal and idealized cross section of the regulated Weissach River.

The actual situation in nature is as follows: Up to now, the bed load material is transported to the Tegernsee and there it will be removed. Quite recently in Germany one tries to restore many small rivers to their so called near-natural situation. At certain reaches big stones were placed in the Weissach River and new ground weirs have been constructed.

Severe sedimentation up to 0.8 m behind the break of the slope appeared after the last of HW events. The decrease of the cross-section area can provocate an overflow of the dikes.

The aim of the model test is to investigate if the near-nature improvement influences the bed load transport. In comparison to the Neumagen River important differences exist here:

a. The Weissach River has no clear geometrically defined cross-sections.
b. There is no possibility to construct a sediment trap on the break of the slope.
c. The regarded river section is limited by a weir to divert a part of the discharge in the head race of a small water power station.

The test procedure in the Weissach River model will be similar to that in the Neumagen River model. The investigation of the sand material was complicated, because lots of grain samples had to be taken to preserve a wide spectrum of grain size distribution (Figure 11).

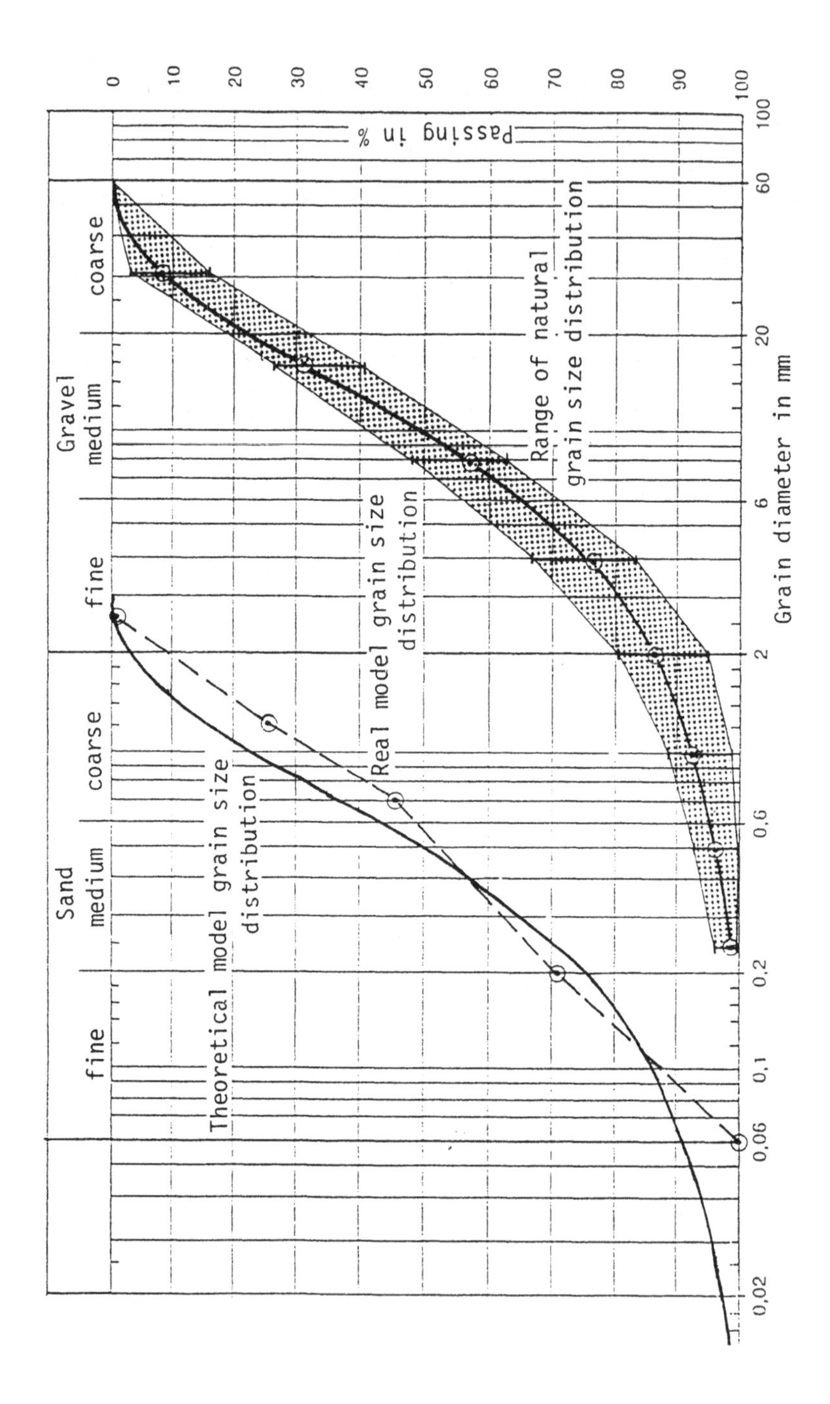

Figure 11. Grain distribution curves of the natural and model bed load material.

The geometrical reduction of the mean grain size curve is also recorded in Figure 11. The choice of the model scale depended on the space available in the LHW. Therefore, the scale number exceeds the calculated number after Günzel. The model material corresponds not exactly to the theoretical grain size distribution. Here we used a commercial quartz-sand, which is produced absolutely clean by the Kaolin-Werk Amberg only.

The hydraulic-morphological calculation showed that during the initiation of motion, the discharge does not take place in the full-rough section ($Re^* \approx 60$). This is going to be reached with a water-depth $h \approx 3$ cm and a discharge $Q \approx 7.4$ ℓ/s = 13 m^3/s (nature). This again corresponds to the demand of a sufficient Weber number even if the current velocities are so high that there is no fear of an influence of the surface tension.

Concerning the discrepancy between the natural situation (e.g. cross sections) and the simplified model, you can neglect the relatively small difference to the full-rough section.

The model consists of two parts for which the slope can be changed independently. The sediment supply takes place by a controllable screw conveyer. A sediment basket had been installed in the model outlet which allows the sand-mass measuring by weighing. The thickness of the sediment layer in the model is nearly
10 cm.

3.2. LIGHTWEIGHT MATERIAL MODEL

At the Federal Institute for Waterways (BAW) in Hamburg the similarity criteria developed by Gehrig were tested the first time in a tidal model of the Elbe River with movable bed (used material: Polystyrol, $\rho_s = 1.05$ g/cm^3). The development of the model and measuring technique demanded a big loss of time. The interpretation of the results had been also very difficult. Looking over a tidal area this is not astonishing for the great dimensions of such a river (e.g. depth and width) and the "simultaneous effect" complicate the analysis of measurements "in situ". There are different publications concerning the model tests. Therefore, only important general remarks concerning the model and measuring technique shall be given.

Movable bed tidal models are often constructed for use over several years. The model material must be such that no breakdown or changes in the material occur. It is necessary to compromise on exact similarity, since normally available material does not satisfy the entire density range of required materials based on scaling laws. The use of polystyrene has been generally found to be applicable for movable bed tidal models. It is especially important to consider the hydrophobic characteristics of the material since this can cause problems during the periodic drainage of tidal flats. Up to this time it has not been possible to avoid the addition of a wetting agent to the water used in the model.

It is not feasible to simulate the natural particle diameter spectrum of the prototype material in the tidal zone. It is useful to use a constant particle diameter for the sake of reproducibility.

Movable bed models should be constructed in enclosed, heatable halls in order to avoid large water temperature variations. It is recommended that the simulated estuary encompass the entire tidally influenced region. Technical simplifications are possible in cases where models with both fixed and movable bed regions are to be used.

The flow distribution in the typically very wide ocean end of the tidal model can be correlated with the existing morphology using a series of individually controllable valved pipes. The control flap must have a tight seal in order to allow maintenance of a constant water surface elevation relative to which changes in the bottom topography can be measured.

The high costs associated with granular synthetic particles compels one to use them sparingly. The model basin must, therefore, essentially conform to the prototype topography in order to allow the use of a uniformly thick "sediment" layer (Figure 12).

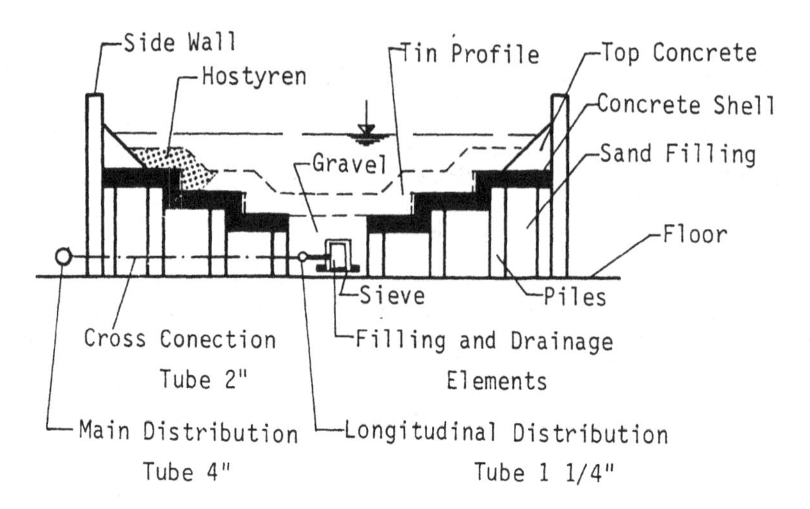

Figure 12. Model cross-section for the movable bed part (schematical).

The use of either the inflow end of the model to fill the model or the control flap to empty the model is not recommended because of the low density of the model sediment (e.g. polystyrene). This would result in a redistribution of the polystyrene. For this reason it is necessary to have an area-distributed system for filling and emptying the model (Figure 13).

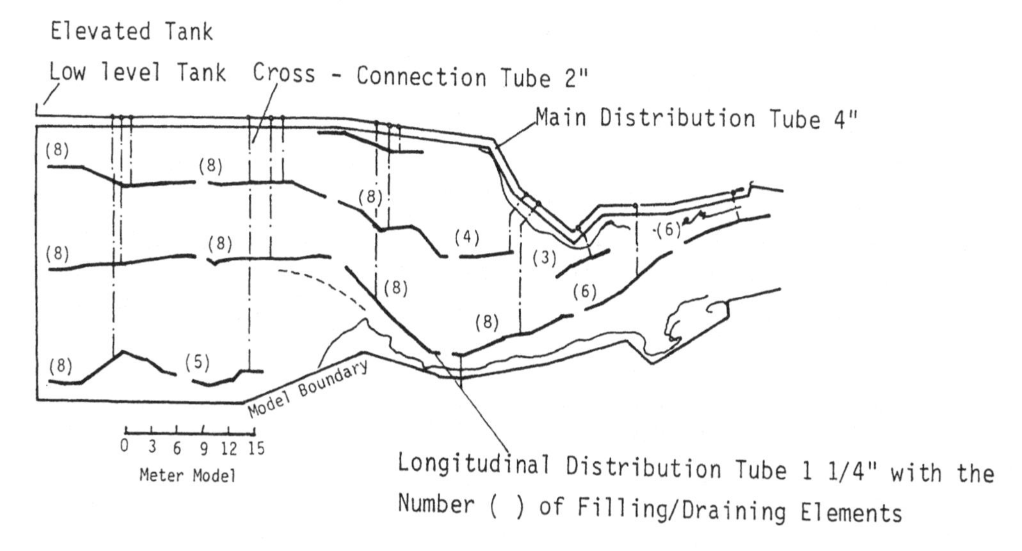

Figure 13. Distribution system for irrigation and draining.

The reproduction of historical hydraulic conditions, as well as the measurement and storage of data are extremely important in small scale movable bed models. Because the slope stability of artificial sediment is not very high topographical measurements must be conducted under water. Acoustical, electrical and optical techniques can be used. The use of acoustical techniques is restricted by the shallow water depths in the model; on the other hand, both electrical and optical measurement techniques can be used effectively. Optical methods have additional advantages if both a movable and fixed bed are present in a given model section. The basic components of an optical bathymeter are illustrated in Figure 14. A constant water level is generally used as the reference elevation because the model width is usually large. The instrument is located on a float and is transported along a prescribed transect using a built-in motor.

Measurement of the bottom profiles alone is not sufficient, however, when conclusions about the sediment flux and transport direction are to be made. One can, therefore, make additional tracer measurements in the model using an appropriate marking of the model sediment. In choosing the marking technique, care should be taken to avoid changes in the hydraulic characteristics of the marked material; the material should also be traceable when it is buried; the markation should not be permanent so that many experiments can be conducted sequentially with the same bottom material. These conditions can be met by using short lived γ-isotopes to mark the sediment. Specially treated polystyrene particles are introduced into the model sediment and thus used as a particle tracer. Since pure polystyrene $(C_8H_8)_n$ is mostly used, radionuclides are not to be expected as a result of neutron bombardment. Elements which are transformed into radioactive isotopes through neutron bombardment must, therefore, be added to the material or to its surface in order to produce the characteristics necessary for the experiment. The half life must be so chosen that, on the one hand, a sufficient time is available to carry out the experiments while, on the other hand, the time is not so long that the radiation intensity requires special safety precautions.

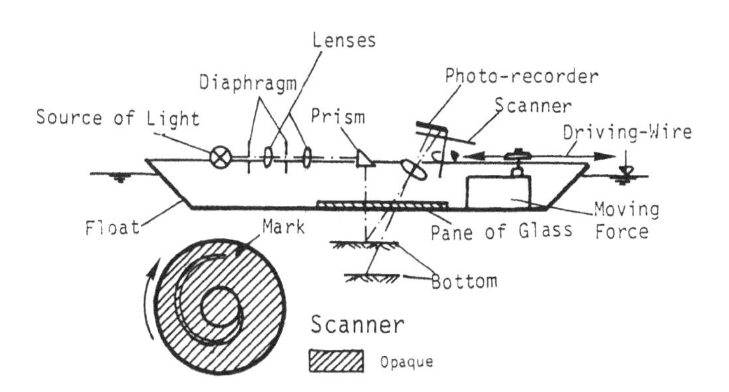

Figure 14. Sketch of an optical bathymeter.

4. CONCLUSIONS

The following remarks should be given:

a. Sand models deliver good results which one can transfer to nature with a great security, assuming that the geometrical reduction is possible and the condition $Re^* \geq 70$ is regarded. Such cases are very rare. On the other hand the relatively small scale numbers are often difficult to realize for the large space they require.

b. Distortion demands the use of model material with $\rho_s < 2.65$ g/cm³. The density graduation of such material is restricted, therefore it is necessary to make compromises. That means the calculation of the scale numbers is an approximation for the practical realization of the model. Space, available material, measuring devices, etc., are of great importance. Unlimited distortion is not possible, sometimes "Froude manipulation" is necessary.

c. It would be desirable that the laboratories have the courage to report about the failures in handling movable bed models. We all can learn about this! In general more should be published about movable bed models. The application of technique and results should be understandable.

5. REFERENCES

Gehrig, W. (1967) *Über die Frage der naturähnlichen Nachbildung der Feststoffbewegung in Modellen*, Franzius-Institut, University of Hannover, Heft 29.

Guenzel, W. (1964), *Modelle geschiebeführender Flüsse mit hydraulisch rauher Sohle*, Theodor-Rehbock-Flußbaulaboratorium University of Karlsruhe, 150. Arbeit.

Vollmers, H.-J. (1986) *Physical Modelling of Sediment Transport in Coastal Models*, Proc. Third International Symposium on River Sedimentation, Jackson, Mississippi, 472–487.

THE PRACTICE OF THE CHATOU LABORATORY IN MOVABLE BED MODELLING

GERARD NICOLLET
Deputy Director of Laboratoire National d'Hydraulique
6, quai Watier
B.P. 49
78401 Chatou
France

ABSTRACT. The paper presents the modelling techniques related to the non-maritime physical models. The most frequent cases to study consists of rivers with bedload and dunes transport. The 25 years' practice has restricted the bed materials to three types only: sand, bakelite and a polystiren. The model laws are presented followed by several examples of application to sand and gravel bed rivers.

1. INTRODUCTION

The present paper explains the practice of the Chatou Laboratory modelling bed-load in rivers (problems of suspension are not considered in this presentation).

Since 1960, the Laboratoire National d'Hydraulique (LNH) has restricted the movable bed materials to three types, which are easily available: sand, bakelite ($\rho s/\rho \simeq 1.4$ and a polystiren called styvaren ($\rho s/\rho \simeq 1.04$). After many trials we abandoned materials like apricot-stone or pumice-stone mixed with sand. Notice that the two lighter materials are restricted in size and shape, due to the process of manufacturing:

- the styvaren particles obtained by extrusion appear like small pieces of "spaghettis". They are cylinders with equal height and base diameter. The LNH has nowadays a big volume of particles of 2 mm and 3 mm, which were treated by a mechanical process to avoid fastening of air bubbles on the particles.
- the bakelite, obtained by grinding old telephone sets, doesn't give rounded particles, but small plates, so that it is not possible to get particles whose characteristic size is larger than 3.5 mm.

The most common cases of bed-load transport are concerned with dunes in natural rivers. Our practice in modelling is based on numerous flume tests on sand waves, whose results are summarized hereafter.

H. W. Shen (ed.), Movable Bed Physical Models, 141–147.

2. CHATOU RESULTS ON SAND-WAVES

The first study, by Chauvin, gave a very simple criterion about the absence of ripples: the dimensionless variable D^* has to be greater than 15 to have a transport by dunes only:

$$D^* \geq 15. \tag{1}$$

The purpose of the second study, by Jensen and Lebreton, was to establish empirical laws correlating the height, the length and the friction coefficient of dunes with the most significant parameters issued from the dimensional analysis. The best-fit relations are the following:

$$\text{wave height} \qquad \frac{\Delta}{H} = 0.85 \qquad Y^{0.68} \tag{2}$$

$$\text{wave length} \qquad \frac{\gamma}{Dm} = 30 \qquad Y^{-0.43} \quad Z^{0.47} \tag{3}$$

$$\text{friction coefficient} \qquad f = 0.27 \qquad Y^{0.64} \tag{4}$$

3. SCALING OF UNDISTORTED MODELS

3.1. Basic Laws for Scaling

- Froude law

$$F = 1 \rightarrow V = \hat{H}^{1/2} \rightarrow Q = V L \hat{H} \tag{5}$$

- Bed-load: we suppose that, in nature, the grain Reynolds number is large enough, so that the bed-load is completely described by the knowledge of the mobility number Y. In such a condition the only law to satisfy is:

$$Y = 1 \rightarrow (\hat{\ })\gamma s Dm = H^2 L^{-1} \tag{6}$$

- Roughness condition: when the bed-load presents sand-waves, it is obvious that the grain roughness is negligible compared to the roughness of the dunes. The only relationship to be considered then is the one in Equation 4:

$$f = \frac{8J}{F^2} = 0.27 \qquad Y^{0.64} \tag{7}$$

If F = Y = 1 the friction coefficient will automatically be correctly scaled and the slope scale J =1.

- Sand-waves geometry: Y = 1 implies that the wave height will be correctly represented on the model, whereas the wave length-relationship (3) – can be written as follows:

$$(\hat{\ })\frac{\lambda}{H} = \gamma_s^{-0.53} \frac{Dm^{0.47}}{D_{90}} \tag{8}$$

This is the only relationship containing the particle size D_{90}.

3.2. Feasability of Scaling

The main limitation comes from the existence of a sufficient level of turbulence in the model to avoid the development of ripples: the conditions (1) applied to Dm gives limit in the model. Then, from relationship (6) we can deduce the lower limit for the grain size Dm in nature, which can be represented without ripples for the considered material.

Bed Material	γ s/g	Dm limit model (mm)	γ s	Dm limit nature (mm)
Sand	1.6	-.6	1	$0.6\ H^2L^{-1}$
Bakelite	0.4	1	4	$H^2L^{-1}/4$
Styvaren	0.04	2	40	$H^2L^{-1}/40$

Figure 1 presents a diagram of the three types of models, where different example of studies performed in Chatou Laboratory are indicated:

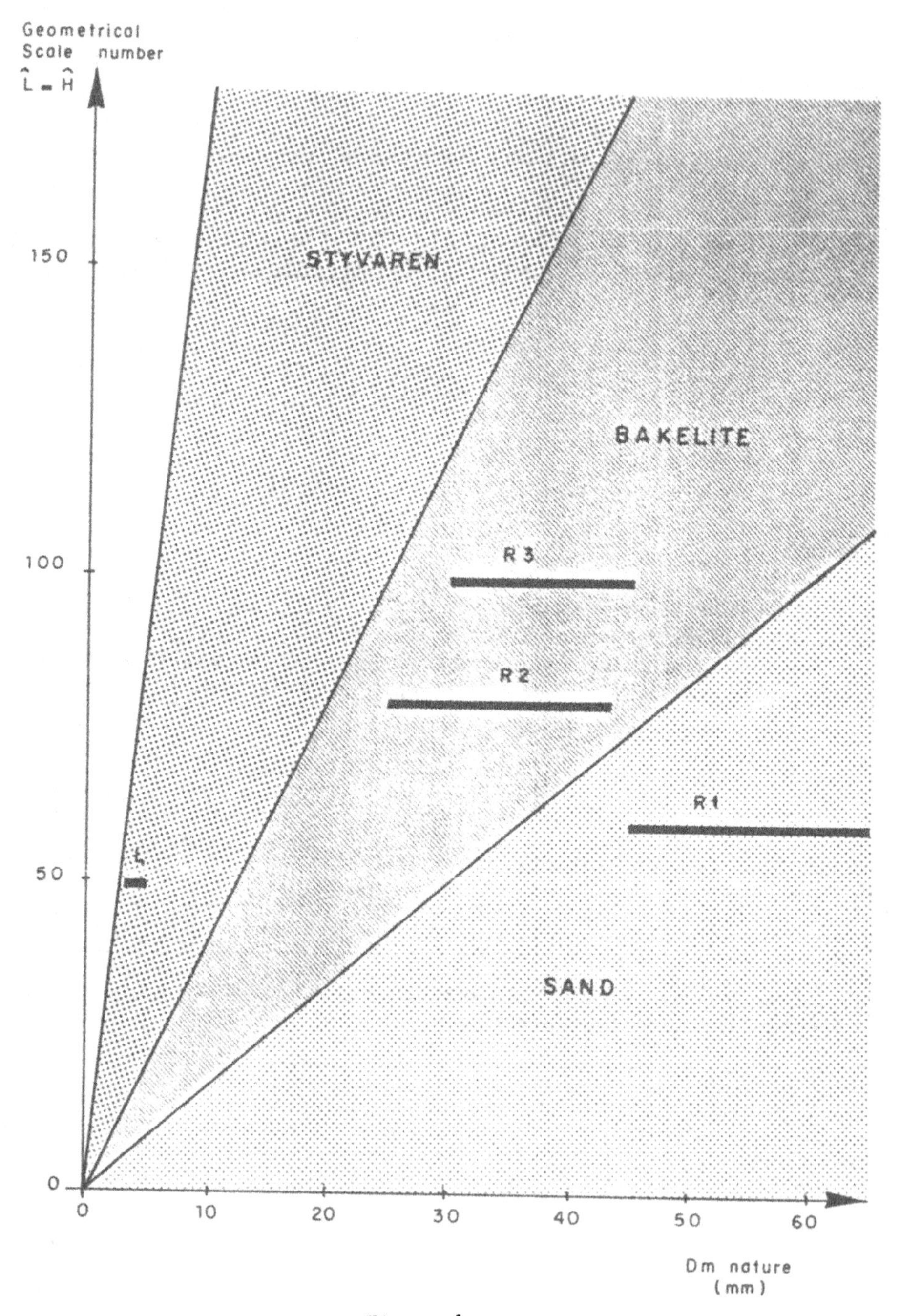
Geometrical
Scale number
L = H
150
100
50
0
STYVAREN
BAKELITE
R3
R2
R1
SAND
0
10
20
30
40
50
60
Dm nature
(mm)

Figure 1.

- R1: This is the first generation of the Rhône River studies performed in the 1960's, with a geometrical scale (S) = 60. The accuracy of measurements is very good, the sand-waves are well simulated, but the models are expensive and the time scale leads to very long test durations (L = 60 gives ts = 7.75).
- R2 and R3: The second and third generations of the Rhône models performed nowadays with bakelite give less expensive models and more reasonable tests durations:

$$(1)\ ts - \gamma s L^{5/2} H^{-2} \qquad (8)$$

For bakelite $\gamma s = 4$ and L = H = 100 ts = 40

The only difficulty is to respect the relationship (7), so that in many cases the dunes length is not well simulated. On the other hand the dunes height and the friction coefficient are correct. The main disadvantage is that the light bed-material increases the local scour near structures.

- L: The domain of rivers with rather fine sand like the River Loire in France. Several models have been used recently for studying the water intake of nuclear power plants on the Loire.

Although the dunes length cannot be represented correctly, the dune height is at the geometrical scale and the headloss is correct.

Notice that the model of St. Laurent des Eaux on the Loire gave good results, whereas a first model 3 times distorted was completely wrong.

4. SCALING OF DISTORTED MODELS

The diagram on Figure 1 shows that the simulation of sands below 2 mm is not possible at a reasonable scale. The distortion is compulsory if Dm nature < 2 mm.

The respect of F = Y = 1 to have correct flow, sediment rate, and dune height on the model, makes the respect of relationship (4) difficult:

$$(^*)f - \frac{8J}{F^2} - 0.27 \qquad Y^{0.64} \qquad (4)$$

It is obviously impossible to satisfy simultaneously $J - \frac{H}{L} \neq 1$ *and* $f - 1$.

However, several models were built for big African rivers using styvaren to represent fine sand, the basic relationship being F = Y = 1 and the distortion less than 3.

As an example, let us consider a river with a sand of size Dm nature = 1 mm and let us try to represent it with styvaren ($\gamma s = 40$). If we choose on the model a bed material of uniform size 2 mm, the sediment scale is Dm = 1/2. If we impose the maximal possible distortion $\delta - \frac{L}{H} - 3$ and the respect of Y = 1, the relationship (6) gives:

$$(\hat{\ })\gamma s Dm - H^2 L^{-1} \text{ with } \hat{\gamma} s - 40, \hat{D}m - 1/2$$
$$\hat{L} - 180 \text{ and } \hat{H} - 60$$

One of the major difficulties comes from the sedimentological time scale (^) ts = γs L5/2 H^{-2} = 4830, which means that 1 hour on the model represents 200 days in nature! Such a nonsteady effect is very harmful to the reliability of the model, as well as the exageration of local scour.

In these conditions, it is wise to consider such a model as a qualitative tool only.

5. CONCLUSION

Physical undistorted models give a good quantitative simulation of bed-load with dunes transport when the grain size in nature is beyond 2 mm. The coarser the sediments are, the better the simulation is.

For sediment below 2 mm it is necessary to build a distorted model, but the accuracy of modelling is restricted and it is recommended not to focus on local phenomenons, but only to consider the general tendencies.

List of symbols:

ν = kinematic viscosity of fluid

ρ = density of fluid

ρs = grain density

γs = (ρs - ρ)g specific weight of a grain in water

D = grain size (Dm = mean diameter, D_{90} = diameter of big elements)

V = flow velocity

Q = flow rate

L = length

H = height

J = H/L = slope of energy line

f = $8J/F^2$ friction factor

(^) = scale such as L, H, V, ...

t = sedimentological time-scale

F^2 = V^2/gH = Froude number

X = $\frac{Dm\sqrt{gHJ}}{\nu}$ = grain size Reynolds number

Y = $\rho\ gHJ/\gamma s\ Dm$ = mobility number

Z = H/D_{90}

D^{3*} = $\gamma s D^3 m/\rho\nu^2$

SCALING OF SEDIMENT TRANSPORT PHENOMENA IN LARGE ALLUVIAL RIVERS WITH VERY LOW SLOPES

J. J. PETERS
Ministry of Public Works
Hydraulic Research Laboratory
Rue de l'Abattoir, 164, B-6071
Châtelet, Belgium

ABSTRACT. A study of the navigation conditions of the Zaire inner delta, initiated in 1968 at the Hydraulic Research Laboratory-Borgerhout, included field observations and model tests. Various model conditions were tried out for the reproduction of morphologic processes: different bed materials, different distortions, different velocity scales, model tilting. After a short description of the experience gained during these investigations, some ideas, thoughts, and doubts about the scaling procedures are presented, as two new models will be designed in the near future.

RÉSUMÉ. En 1968 fût entreprise au Laboratoire de Recherches Hydraliques-Borgerhout une étude des conditions de navigation dans le delta intérieur du fleuve Zaïre. Cette étude devait comporter des études en nature et des essais en modèles ont été testées en vue de la reproduction des processus morphologiques: différents types de matériaux de fond, différentes distorsions, différentes échelles de vitesses, l'inclinaison du modèle. Après une brève description de l'experience obtenue pendant ces essais, quelques idées, réflexions et doutes sont présentés au sujet des procédures de détermination des échelles, aussi en vue de la construction future de deux nouveauz modèles.

1. INTRODUCTION

The Zaire River presents over the major part of its course very low slopes and a sediment transport predominantly as bedload of medium size sand. In the maritime reach, an inner delta was formed there where the river leaves the Mayumbe Mountains, filling up the large submarine canyon cut out in an old continental shelf.

The delta ends at the head of the salt-wedge estuary, thirty kilometers upstream the coastal line. In this sedimentation area, 60 km long and over 10 km wide, a complex system of islands and shoals was developed, in which it is tried to maintain a 28 feet deep navigation channel by dredging. The slopes of the energy lines range from 1 to 10 cm per km, rising locally to 20 cm per km in exceptional cases.

H. W. Shen (ed.), Movable Bed Physical Models, 149–158.

The Hydraulic Research Laboratory of the Belgian Ministry of Publics Works was entrusted in 1968 by the Belgian Ministry of Foreign Affairs and Technical Cooperation with a study of the improvement of the navigation channel by dredging operations and the optimization of these.

A partial model was initially built. It reproduced the area of a new channel opened for navigation in 1968. Field observations were performed at that time in order to provide data for its calibration. They revealed various peculiar characteristics of the sedimentologic behavior of the inner delta:

- selective sedimentation producing a sorting of the bed material, the sizes varying from place to place;
- change of bedform types during floods, related among others to the above mentioned spatial size distribution;
- complex meandersystem whose evolutions are mainly controlled by geologic and hydraulic conditions, the banks eroding generally very easily;
- large cross-sections in which the flow patterns are affected by spiral flow and lateral exchange of momentum and turbulent energy;
- spatial variation of bedform type and of hydraulic roughness;
- the rate of sediment transport is not only related to local flow conditions.

Due to these statements, it was decided to combine specific model tests and extensive field observations, in order to improve our understanding of the dynamic behavior of the river. The database established until now allowed a better insight in the sedimentologic characteristics of the river reach, but raised a number of new questions.

In order to solve these, two new models are planned, each of them with specific aims:

- the first to study the effects of geologic controls in the upstream stretch, this control determining the solid discharge distribution (amount and size) entering in the complex meandersystem;
- the second to analyze the evolutions of this meandersystem, considering the best way to intervene in its process, especially by dredging.

2. PREVIOUS MODEL TESTS

2.1. Distorted Model

The model reproduced the area of a new meandering navigation channel opened in 1968, only one main arm of the delta. At the boundaries, the flows to other river branches or entering the model at confluences were controlled by a system of weirs and pumps (Figure 1). The major scaling criteria were:

a. For the fixed bed model:

- distortion factor of 5, because of the available room (vertical scale factor nh = 500) and because of the very low water surface slopes (horizontal scale factor nl = 100);

- Froude condition for the flow (velocity scale factor nv = 10);
- reproduction of sediment transport paths over a fixed bed with either bakelite or polystyrene particles.

b. For the movable bed model:

Horizontal and vertical scales were the same as for the fixed bed model.
The bed morphology changes were investigated only in part of the model. The upper and lower parts of the model, as well as some more stable shoals, were made as fixed bed areas. The bed of a meander, including the concave banks, was reproduced with movable bed material (Figure 1).
The bed material and velocity scales were selected as follows:

- selection of the bed material with flume tests on bakelite and polystyrene for the best fitting of the roughness; bakelite was chosen;
- adjustment of the velocity scale for initial scour (nv = 8.33 different from Froude condition nv = 10) in the case where bakelite was used as bed material.

The major problems encountered during the tests were related to instability of the concave banks, mainly when using polystyrene. It was tried to stabilize steep sloping banks using a mixture of the particles of the bed material with cement. The cement content obviously appeared to be very critical: it had to be high enough to prevent bank sliding but low enough to not inhibit bank erosion.
Bakelite was best reproducing the morphologic changes. Polystyrene bedforms were too large (too smooth) and the concave bank slopes too flat (Figure 2).
Considering all the modeling problems due to the distortion, it was finally decided to modify the model scales.

2.2. Tilted Model

It was decided to try the tilted model with bakelite as bed material. The scaling procedure is well known:

$$-n_s = n_D^{3/2} \cdot n^{1/2}$$

$$-n_v = n_D^{1/2} \cdot n_\Delta \cdot n_{C90}^{3/4} \cdot n_C^{1/4}$$

$$-n_C^2 = n_l/n_h$$

$$-\delta = I_{pr}\frac{(n_C^2 \cdot n_{h-n_l})}{n_v^2 n_h}$$

s = transport per unit time and unit width

D = average sediment size

C = Chezy coefficient

Δ = $(\rho_s - \rho_w)/\rho_w$

I = slope

δ = tilting angle

The calculated velocity scale factor n_v = 4.5" induced too large transport rates, and it was decided to use the following model scales:

horizontal scale factor	n_l	=	500
vertical scale factor	n_h	=	166
velocity scale factor	n_v	=	7
tilting angle	δ	=	2*10E-4

The critical elements in this procedure appeared to be the transport formula, the determination of roughness of the bed material--both in the model and in the prototype--and also the choice of the distortion factor.

The reach reproduced by the model was almost the same as for the first model, hydraulic structures controlling the flow field at places of separation of flow and at confluences. The floods being very progressive ones--floods with almost a doubling of the discharge in four months--the hydrogram injected in the model was a succession of four different, constant discharges. The sedimentologic time scale factor was not computed theoretically but determined experimentally on the base of historical calibration.

Because of the hydraulic restraints (discharge and levels regulated at all inputs and outputs of the model) the sedimentologic behavior was easier to control. Furthermore, the deviation of the general flow direction from the longitudinal axis of the model was limited and there did not appear marked scale effects due to the tilting. It was, however, necessary to start up with an initial roughness making dunes by hand in the movable bed.

For different circumstances the model investigations had to be interrupted and it was unfortunately not possible to gain more information about the sedimentologic behavior of the tilted model. Nevertheless, the general sedimentological behavior of the model appeared to be quite good.

3. FUTURE MODEL TESTS

In the meantime our knowledge of the area under investigation improved and allowed us to identify the most important processes to be reproduced in the models. These were subdivided in two groups. The first, occurring in the upper reach of the delta, is related to the sediment segregation, due to selective sedimentation and secondary flow circulation. The second is related to the meander development in the complex island system in the lower part of the area. Each will be investigated with a proper model.

3.1. Model Upper Reach

In the area to be reproduced, the banks are mostly stable, formed of rock or clay. The main effect of floods consists of the storage in some stretches of large amounts of finer sands on shoals whose positions are determined by the geologic and hydraulic controls. The heterogeneousness in space of the size of the bed material due to selective segregation is very important in some places. Sediment sizes may range from hundred to thousands of microns in a two kilometer wide cross-section. The magnitude of the floods, resulting in the flow currents as well as in the water levels, regulates the mobilization of these sediments, controlling the meander development in the lower reach. This means that the history of hydrological cycles is of utmost importance in the study of the evolution of the navigation channel.

The aim of the model is to study which secondary currents control the shoaling processes and to investigate the possibility of restraining them by river works.

3.1.1. *Scaling Criteria* Because of the importance of the sorting of sediment sizes, it seems necessary to use a model-sediment with a marked size distribution. Sand seems to be the only solution, according to the size of the model and the cost of the necessary amount of bed material.

Distortion should be limited because of the necessity to reproduce the helical flow in some marked bends and to prevent bank sliding. The energy line slopes being of an order of magnitude between 10E-4 to 10E-5, the difficulty resides in avoiding model tilting.

In some areas, however, the principal direction of the flow is South-North, perpendicular to the overall East-West direction of the river. The slope of the energy grade line would thus not be magnified with the same ratio in all the channels. The question is to know which is the most important model criterion: reproduction of the helical flow pattern or the energy slope in the different branches.

We are looking for a solution in order to adjust the relative sediment transport capacities in distinct arms. This implies a design of the cross-sections realizing a compromise between transport capacities of both water and sediment, supplying if necessary some additional energy. This could be performed, for example, by means of paddle wheels or water ejectors (procedure somehow similar to the correction of coriolis forces). Flows between river arms through inundation areas at extreme flood conditions, could probably be controlled.

Another important criterion will be the sediment sorting. Field measurements have shown the role played by the depth in the transport mechanisms. Local shear stresses seem not to be related univocally to the energy line slopes. Shear stresses computed with the depth and energy line slope may differ strongly from those deduced from the measured velocity profiles.

$$(\tau_0)_1 - \gamma R I e$$

$$(\tau_0)_2 - \rho V_*$$

τ_0 = bed shear stress

γ = density

ρ = specific mass

V_* = shear velocity

R = hydraulic radius

The differences of bedform types encountered simultaneously in the river course and their changes during floods will not be taken into account in the model.

The following procedure for determining the model scales is proposed:

- neglect the Froude condition;
- determine a first set of distortion ratio and tilting angle on the base of the selected (available) model bed materials, bakelite or sand. The tilting angle is defined along the thalweg taken in the main channel;
- investigate in test flumes the relationship between the transport parameter $\Psi - s/\sqrt{D^3 g \Delta}$ and the flow parameter $\Theta - \tau_0/\rho g \Delta D$ for the different available model bed materials. It will be compared with the relationship obtained from field data. Special attention will be paid to the difference observed in the prototype between $(\tau_o)_1$ and $(\tau_o)_2$;
- readjust the tilting angle and distortion ratio, trying to maintain the latter as close as possible to unity. Sand will preferably be taken as bed material because of the possibility to choose its grain size distribution, but also for its maintenance costs;
- determine the lateral distortion ratio (different transverse and longitudinal length scales) for the important side river arms. This would allow a better reproduction of the distinct transport capacities, for water as well as for solids, in the different river branches.

3.2. Model Lower Reach

This model will be far more difficult because:

- the slopes reduce progressively when approaching the ocean and their local values are strongly affected by bar formations on the shoals (crossings);
- the flow divides into an intricated channel system with numerous branches; the bedform types change with stage depending on the sediment sizes and local flow conditions;
- the flow structure, time averaged as well as turbulent, is complex and strongly three-dimensional; important lateral exchanges exist within and between channels.

A tilted model seems to be advisable because of technical reasons: requirements for precise slope measurements, and use of bakelite or preferably sand. However, the morphologic evolutions are governed by geologic controls and differential bank erosion. Therefore, it is not advisable to build the model with fixed (concrete) banks.

As for the model of the upper reach, it is proposed to investigate the possibility of modifying the transverse length scale of some side arms, fixing their banks temporarily and adding kinetic energy. In this way it could be possible to control liquid and solid discharge distributions.

An important factor seems to be the spatial distribution of sediment transport related to the flood level. The laws of transport rate as a function of depth and velocity will be investigated in test flumes in order to adjust the depth and velocity scales.

4. SUMMARY OF EXPERIENCE WITH THE ZAIRE RIVER MODELS

Tests were performed until now with movable bed models of the very large Zaire River in an area characterized by very low slopes and a complex island system. The scaling procedure, using a small distortion ratio, tilting the model and abandoning the Froude condition for the velocity scale, produced good results when only one branch was modeled. The flow field at the boundaries of the movable bed reach was strongly controlled by a system of weirs and pumps.

Field observations have shown, in some cases, the importance of the sediment sorting due to selective sedimentation, and of changing roughness (bedforms).

Two new models will be built, each with specific aims. In one model, the influence of secondary (helical) flow on the distribution of the sediment transport between different branches, and the consequent shoal formation will be explored.

This tilted model will have stable (concrete) banks and the bed material will preferably be sand. It will be tried to correct the scale effects due to the different influences of the tilting in main and side channels by choosing different transversal length scales in these side channels, adjusted according to the liquid and solid discharge distributions.

A second model will reproduce a reach where geologic and hydraulic controls determine the morphologic changes of the complex alluvial channel system. The model should also be tilted, but it is not yet clear how to reproduce the changes in bedform type observed in the prototype.

5. SUGGESTIONS FOR DISCUSSION

- Importance of sediment size distribution (size distribution of the sediment samples but also the spatial variation of it);
- Procedure to model roughness variations due to changes in bedform type, in particular when passing from lower to upper flow regime (bedform parameter);
- Basic studies about the meaning of bed shear stress in prototype and in model, especially trying to measure them and compare them with the values computed from flow measurements (depth and vertical velocity profile measurements). This must be done in relation with bedform prediction formulas;
- Analysis of procedures to model complex island systems, eventually with overland flow.

6. REFERENCES

Coen, I., Peters, J. J., and Sterling, A. (1968–1977). *Etudes sur modèles réduits, Navigabilité du bief maritime du fleuve Zaïre.* Laboratoire de Recherches Hydrauliques du Ministère des Travaux Publics de Belgique. (Partly non-published.)

De Vries, M. (1973). *Application of physical and mathematical models for river problems.* Delft Hydraulics Laboratory, Publication No. 112.

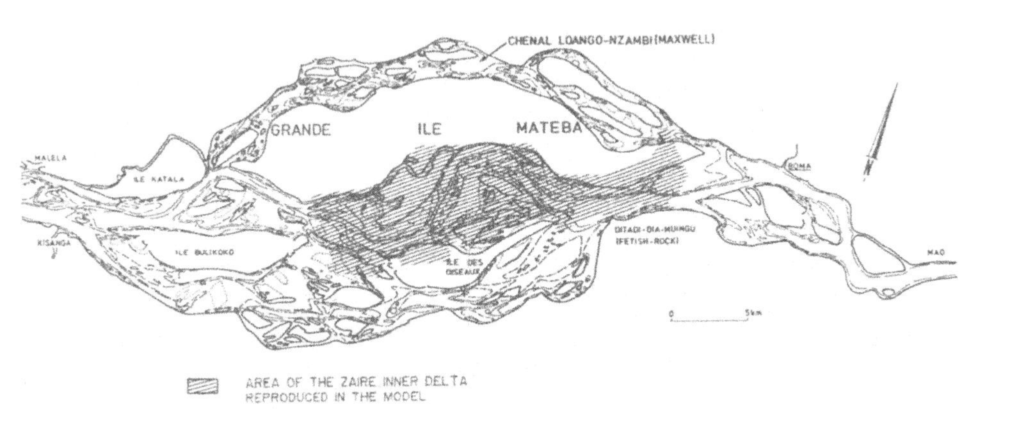

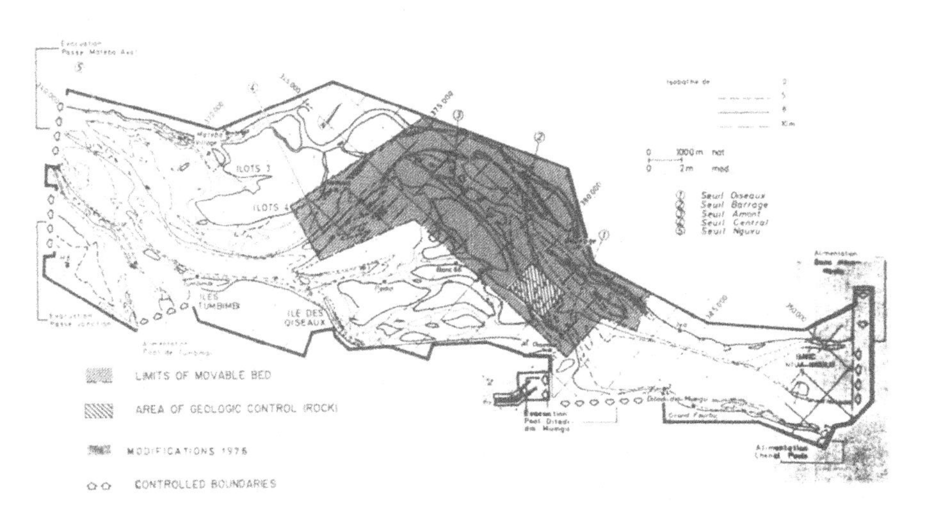

Figure 1. Plan view of the Zaire inner delta and of the model, with indication of the central part transformed for movable bed tests

(a) bakelite

(b) polystyrene

Figure 2. Comparison of movable bed tests with bakelite (a) and polystyrene (b)

MODELING DESILTATION OF RESERVOIRS BY BOTTOM-OUTLET FLUSHING

GIAMPAOLO DI SILVIO
Institute of Hydraulics
University of Padua
Via Loredan 20, 35100 Padova
Italy

ABSTRACT. The problem of preventing the total and irreversible obstruction of the outlet has been examined on the model. Two typical configurations have been considered: (a) short tunnel through the dam, and (b) long tunnel by-passing the dam. After a qualitative description of the flushing operations and of the behaviour of the outlet, the model laws and similitude criteria have been discussed in the paper. Some quantitative results obtained from the model of type (b) are also provided.

1. INTRODUCTION

Whenever favourable circumstances occur, an efficient desiltation procedure for reservoirs may consist in flushing water through the bottom-outlet. Conditions for a substantial removal of material from the reservoir by this method regard topography, sediment characteristics, and availability of sufficient flushing discharge.

In this paper, however, we shall not deal with the problem of an efficient clearing of the reservoir, but rather with the problem of avoiding the total obstruction of the outlet during the operation.

Before discussing the model laws to be applied for simulating desiltation processes, in the following paragraphs a qualitative description will be given of the flushing operation, as well as of the possible causes of clogging.

2. A QUALITATIVE DESCRIPTION OF THE FLUSHING OPERATION

For the sake of example, two typical outlets will be considered, both studied in the model in the laboratory of the Institute of Hydraulics of the University of Padua.

H. W. Shen (ed.), Movable Bed Physical Models, 159–171.

Type A: Short tunnel through the dam, having the downstream-end at the same level of the river bottom;

Type B: Long tunnel by-passing the dam, having the downstream-end at a higher level than the river bottom.

A schematic description of the flushing operation is given in Figures 1 and 2 for the outlet of Type A, and in Figures 3 and 4 for the outlet of Type B. As portrayed in the figures, just after the opening of the gates (Figures 1a and 3a), a funnel-shaped crater ("flushing cone") is created by the flushing water, which removes the material from the immediate vicinity of the outlet. Once the cone has been formed, the water flowing through the tunnel is practically clear until the water level in the reservoir remains high enough.

When the water level comes down to the initial sediment level (Figures 1b and 3b), the "grazing flow" starts to erode the rim of the flushing-cone and to cut a large gully which propagates in the upstream direction. During this phase the amount of material conveyed to the outlet is very large and sediments tend to be deposited wherever the flow tends to slow down. In the first type of outlet (Figure 1b), deposition takes place in the river just downstream the tunnel, while in the second type (Figure 3b), deposition starts in the tunnel itself, where it rapidly reaches the conditions of maximum obstruction.

As bottom erosion in the reservoir propagates upstream (Figures 2c and 4c), the gully becomes less steep and deep, so that the sediment transport toward the outlet progressively decreases. During this phase, however, in the short and steep tunnel (Figure 2c), deposition moves upstream from the river, eventually reaching the conditions of maximum obstruction.

At the end of the flushing operation (Figures 2d and 4d), the sediment transport is strongly reduced and the obstruction of the tunnel has been removed by the flow of clear water.

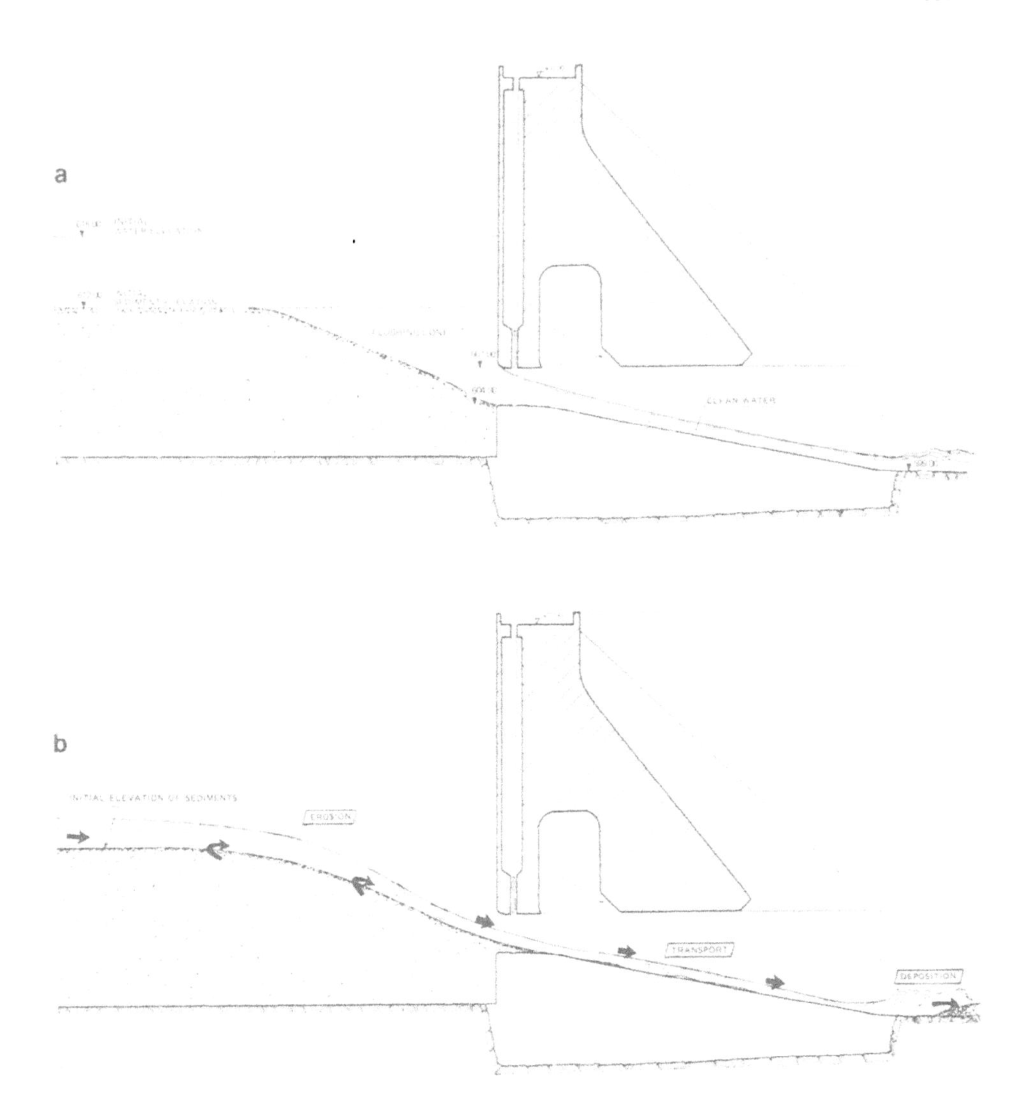

Figure 1. Desiltation process by flushing the bottom outlet. (Type A: Short tunnel). (a) Creation of the "flushing cone" just after the opening of the gates and water flowing through the outlet remains clear until water level in the reservoir is high enough. (b) Erosion of the bottom in the reservoir takes place as soon as "grazing flow" begins; deposition occurs downstream of the outlet.

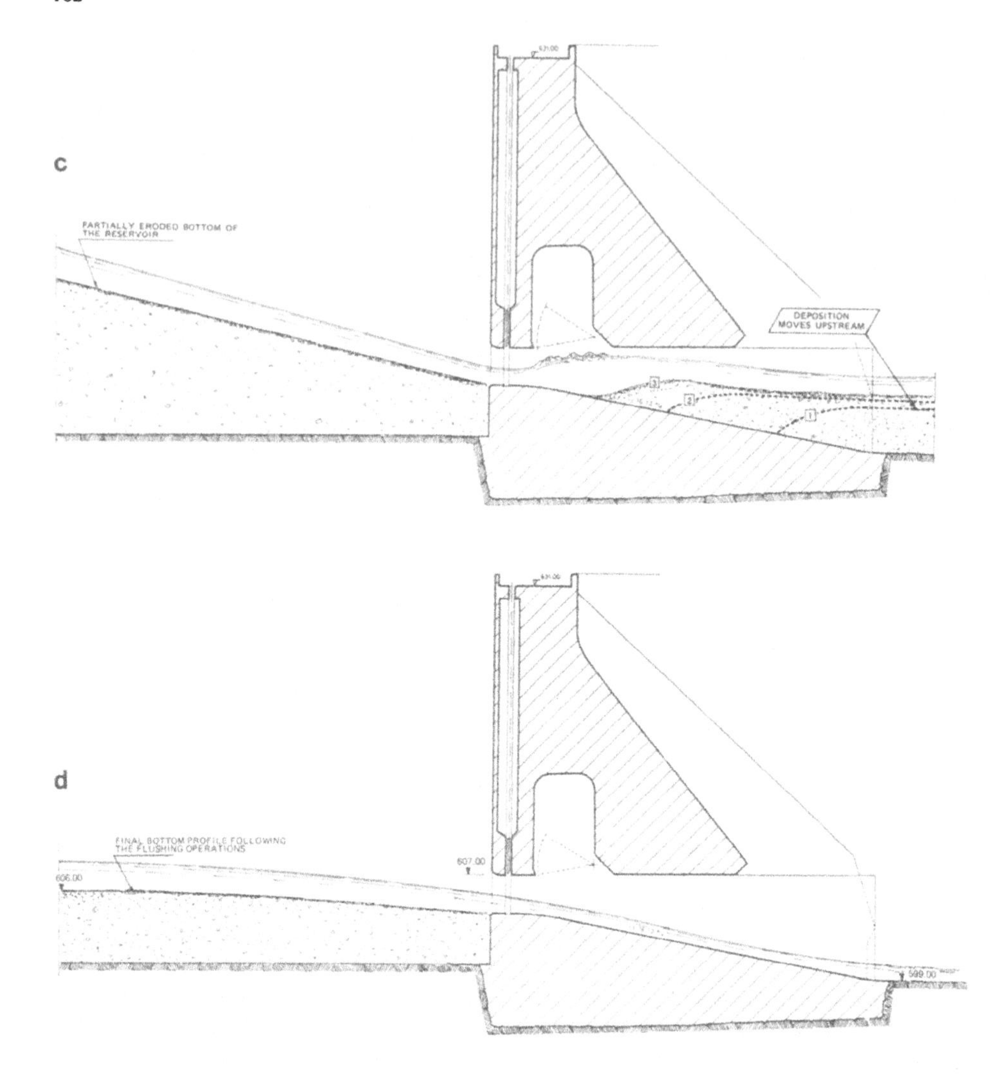

Figure 2. Desiltation process by flushing the bottom outlet (Type A: Short tunnel). Continuation from Figure 1.
(c) Deposition moves from the river into the outlet, while erosion of the bottom proceeds in the reservoir; figure depicts the maximum obstruction of the outlet.
(d) Sediment yield from the reservoir decreases as bottom profile becomes less and less steep; figure depicts the final configuration with deposition completely removed.

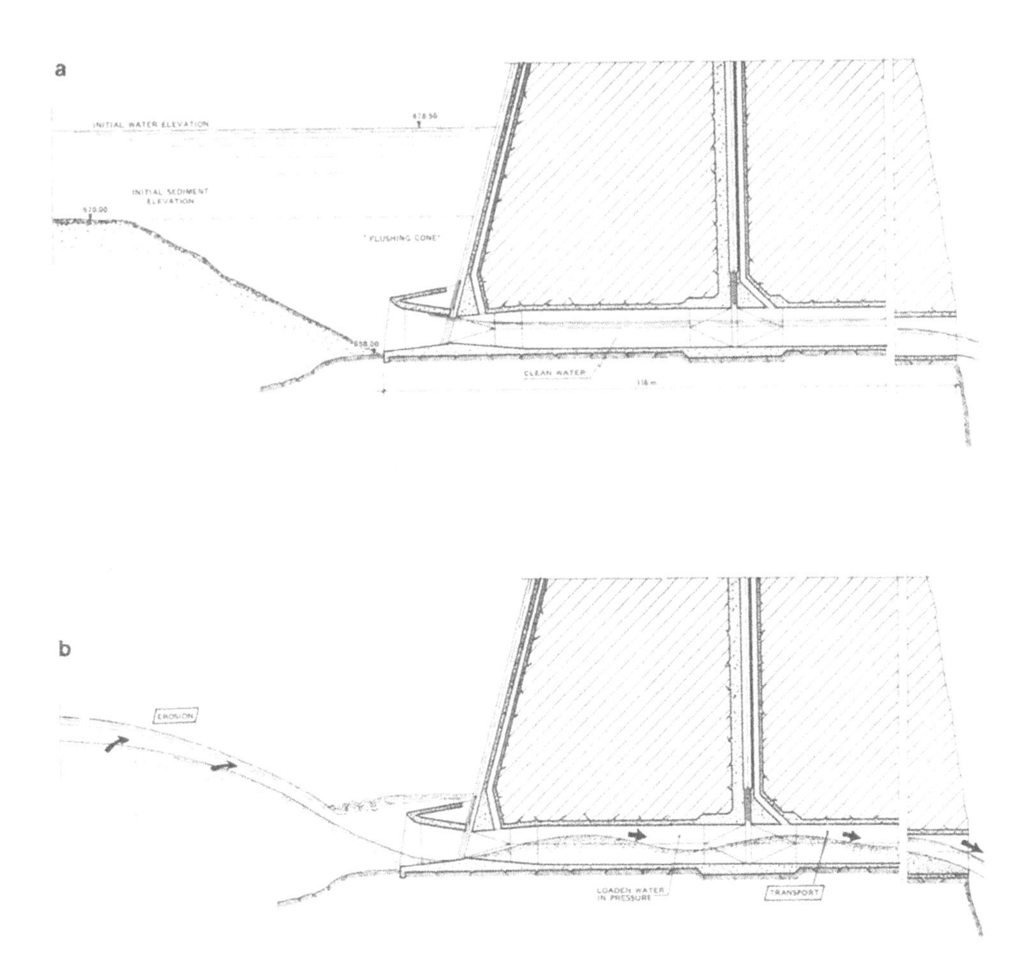

Figure 3. Desiltation process by flushing the bottom outlet (Type B: Long tunnel).
(a) Creation of the "flushing cone" and flowing of clear water;
(b) "Grazing flow" erodes the bottom of the reservoir while deposition takes place in the tunnel; figure depicts the maximum obstruction.

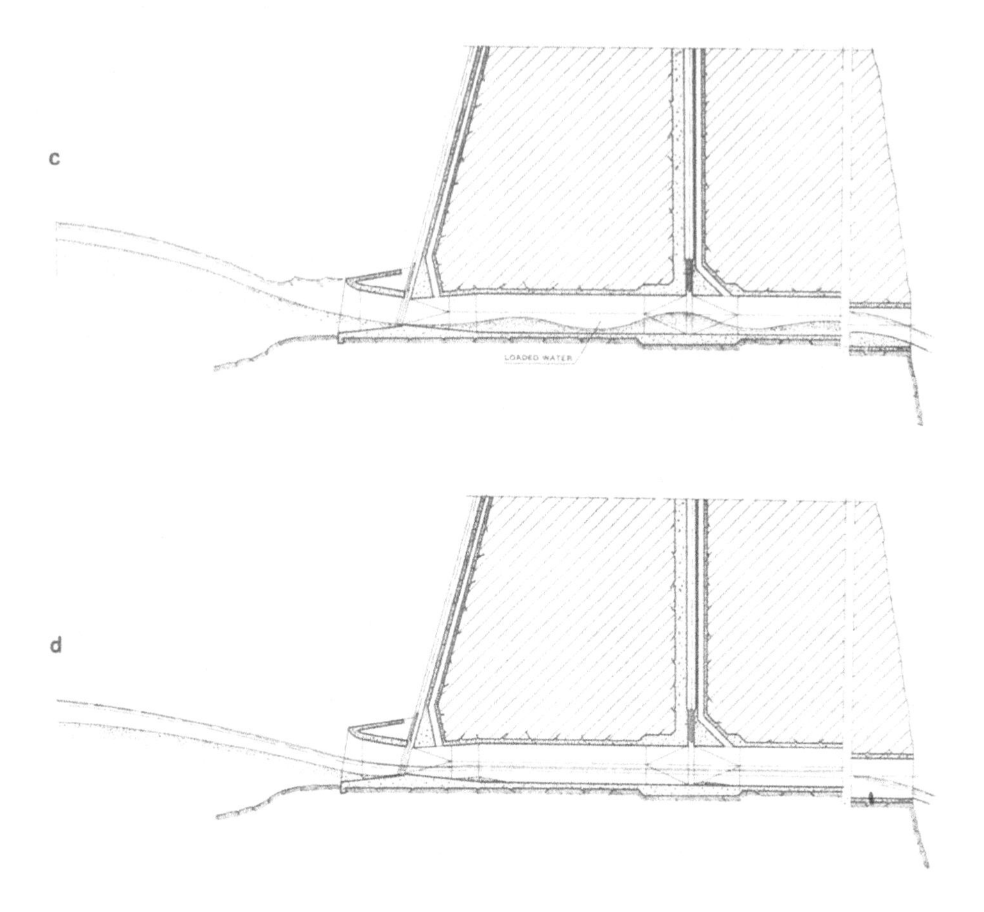

Figure 4. Desiltation process by flushing the bottom outlet (Type B: Long tunnel). Continuation from Figure 3.
(c) Sediment transport from the reservoir decreases as the flushing proceeds;
(d) Configuration at the end of the flushing operation; transport is very small.

3. POSSIBLE CAUSES OF OUTLET CLOGGING

A total obstruction (clogging) of the outlet may occur for several reasons. Figure 5 shows the original configuration of the short tunnel (Type A) having the upstream-end at about the same level of the river bottom: in this condition the outlet becomes completely choked-up by the sediments deposited in the stream, as soon as the flushing discharge is less than 20 m3/s.

Figure 6a shows how the long tunnel (Type B) may be completely clogged due to a sudden interruption of the flushing discharge fed into the reservoir: in fact, while the rapidly decreasing flow is able to bring the eroded material into the outlet, it is not sufficient to convey it through the tunnel, especially if the velocity is reduced by a partial closure of the downstream gate.

Another possible cause of clogging (Figure 6b) is that the initial cross-section is too narrow and the water head too small with respect to the load of sediments weighting on the outlet. In this case, although the seepage can mobilize the material, the water discharge is not suffient to wash it away along the tunnel; consequently, the sediments move downstream as a mixture at very high concentration, until they are stopped by solid friction at a certain distance from the gate.

Occasional clogging of the Type B-tunnel may also take place during the normal flushing operation (see Figure 3b), when the pulsating character of the transport suddenly conveys into the outlet a large bulk of material. These obstructions, however, are soon removed by a temporary increase of the waterlevel in the reservoir.

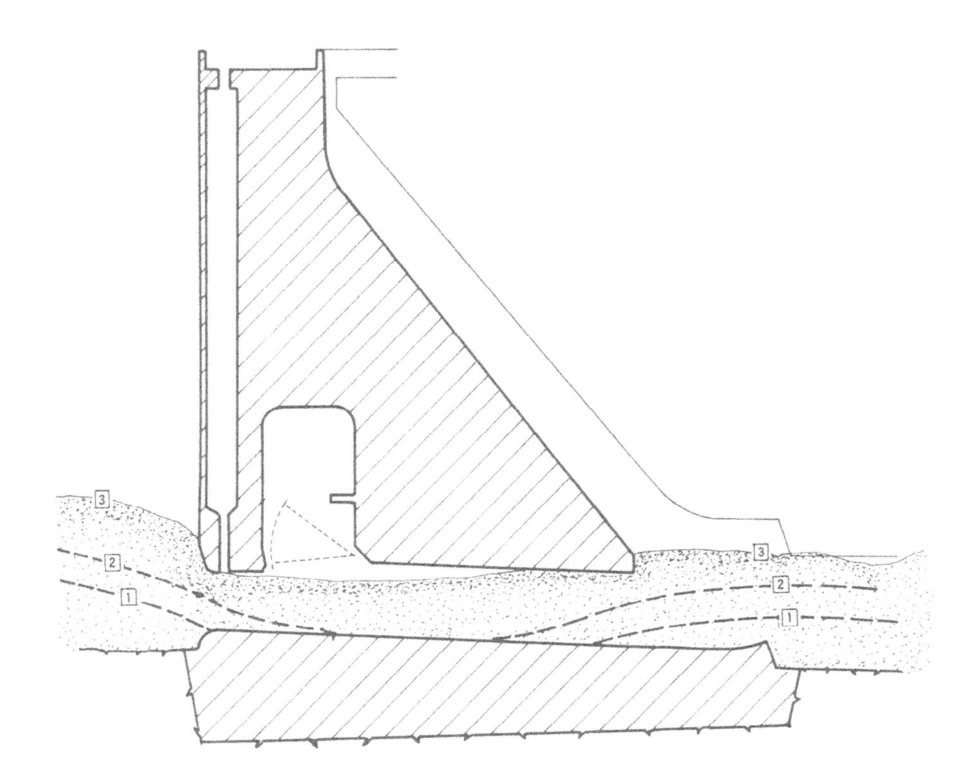

Figure 5. Clogging of the short tunnel (Type B, with a lesser steepness), due to the deposition from the river choking-up the outlet.

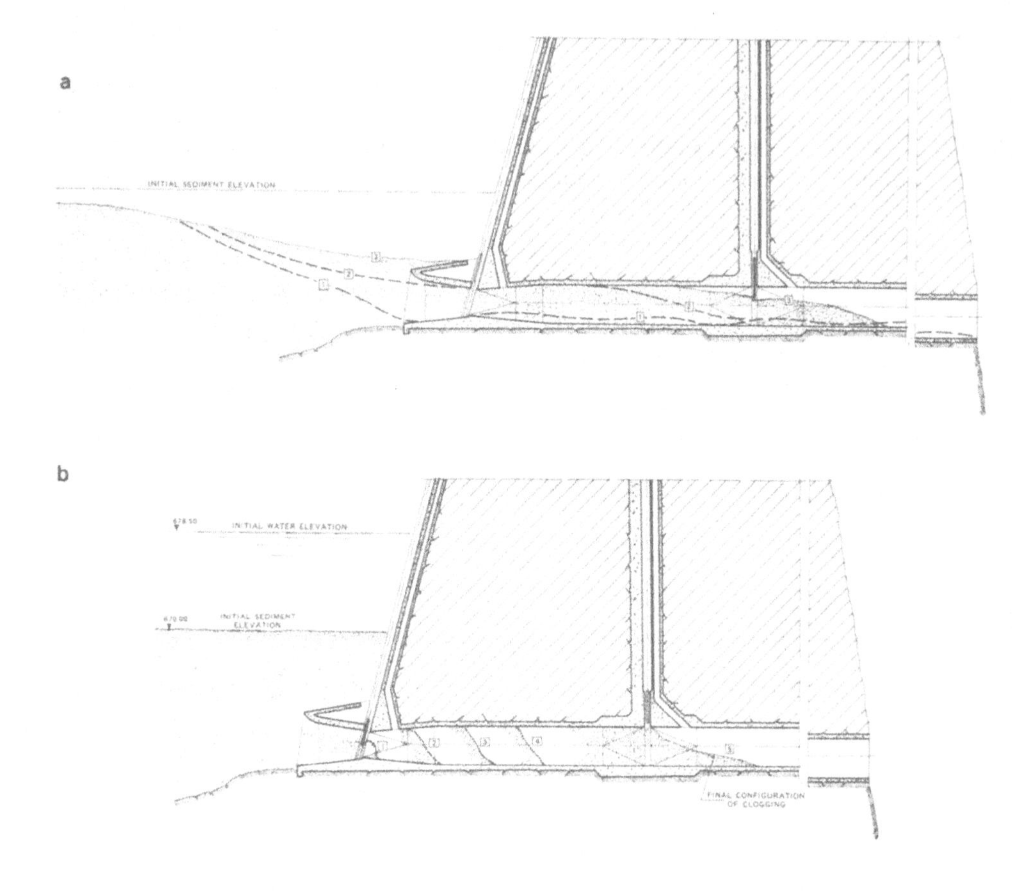

Figure 6. (a) Clogging of the long tunnel (Type B, with partial opening of the downstream gate), due to a sudden interruption of the flushing discharge.
(b) Clogging of the long tunnel (Type B, with partial opening of the upstream gate), due to a too small size of the initial cross-section.

4. SIMILITUDE CRITERIA

The prevailing Froude's criterion conserves the ratio between gravity forces and inertial forces, as well as between gravity forces and solid-friction forces, provided that shape, density, and porosity of the material are the same in model and prototype.

Froude's criterion alone, however, does not guarantee the reproduction of the following aspects of the phenomena, as described in the previous paragraphs:

- Bottom erosion and collapse of banks in the reservoir;
- Transport of sediments along the gully;
- Transport of sediments along the outlet tunnel;
- Deposition of material in the tunnel and in the river downstream;
- Seepage flow and sediment removal by seepage.

Some of these aspects are mutually related, as for example, erosion, transport, and deposition rates; moreover, bank collapses in the reservoir are produced both by foot erosion and by seepage flow. At any rate, all of the above mentioned aspects are basically controlled by the particle size, which should be properly scaled down for model simulation.

Assuming a non-cohesive sediment, the selection of the particle size in the model has been made on the basis of the following criteria: (1) conservation of the ratio between falling velocity and flow velocity; (2) conservation of the ratio between seepage velocity and flow velocity and (3) conservation of the ratio between friction velocity and flow velocity.

In most reservoirs of mountain areas, deposits in the vicinity of the outlet are basically constituted by sand and silt. Within the Stokes' range (d) (1–2 mm) the first criterion (1), associated to the inescapable Froude law, provides:

$$\lambda_d = \lambda^{1/4} \tag{1}$$

where λ_d is the scale of the sediment size and λ the geometrical scale of the model. Note that condition (1) respects as well the second criterion (2) when the seepage velocity is within the Darcy's range.

As for the third criterion (3), the characteristics of the flow, both in the gully and in the tunnel, are definitely those of the upper regime. The "pseudo–bedforms" observed in the model and prototype are in fact produced by repeated collapses of the gully banks. These masses of material are rapidly eroded by the flow, transported in suspension downstream and eventually deposited along the tunnel in the shape of intermittent forms similar to dunes (see Figure 3b and 4c). Therefore, apart from the occasional, localized pseudoforms produced by bank collapses, the actual bottom roughness to be simulated in the model is due to the largest sizes of sediment present on the bed.

Since during previous flushing operations coarse sediments have been transported downstream from the upper part of the reservoir, gravel, cobbles, and even boulders may be found near the outlet. As these coarse elements are well behind the Stokes' range, the third similitude criterium (3) requires:

$$\lambda_d = \lambda \tag{2}$$

In other words, in order to satisfy the above mentioned requirements, the cumulative size–distribution curve of the model should be obtained from the curve of the prototype by applying a scale, gradually varying between Equations 1 and 2, depending on the particle size. In practice, it is sufficient to apply the relevant equation to d_{10} (representative diameter of the seepage flow) to D_{50} (representative diameter of transport) and to d_{90} (representative diameter of bottom roughness).

5. MODEL RESULTS

Some results obtained in the model of the Type–B outlet will be discussed in the present paragraph. The model has a scale of 1:15 and reproduces a part of the reservoir for about 100 m from the outlet. The purpose of the model was to examine the outlet behaviour during the initial (critical) phase of the flushing and to give suggestions for optimizing the operation.

Information on the bed material of the prototype was scanty and far from being accurate; so it was decided to perform a sensitivity analysis on the model with two different grain size distributions: (1) sand, and (2) a mixture of sand (2/3) and gravel (1/3).

The granulometric curves of the two components are given in Figure 7, both for the model and the prototype according to the scales expressed by Equations 1 and 2.

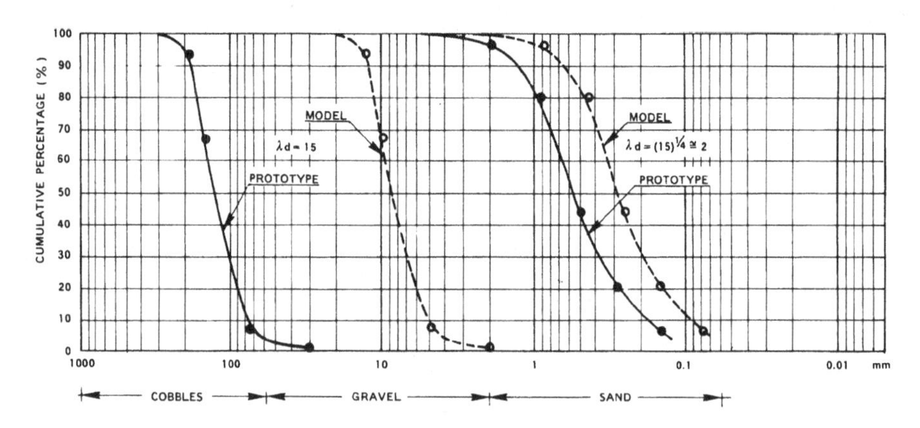

Figure 7. Bottom materials used in the model.

A number of tests (most of them using sand) were made to evaluate the behaviour of the outlet in different conditions:

- initial elevation of the sediments in the reservoir;
- flushing discharge;
- partial closure of the gates.

The following measurements were made on the model during the flushing operation:

- continuous recording of the water level before the outlet;
- survey of the deposits along the tunnel at short time intervals;
- measuring the total amount of sediment discharged through the outlet with different flushing durations.

The most significant quantity was found to be the average concentration:

$$\bar{C} = V_s / V_w \tag{3}$$

defined as the ratio between the volume of sediment V_s, discharged through the outlet during the flushing period, t_f, and the corresponding volume of water, $V_w = Q\, t_f$; note that the duration tf is measured from the beginning of the "grazing flow" (see Figures 1b and 3b), when the gully in the bottom of the reservoir starts developing.

For a given duration of the flushing, the average concentration is basically controlled by the initial elevation of the sediment: more precisely, the value of $\bar{C}$ corresponding to a certain period t_f, is proportional to the initial drop of the waterlevel, Y_o. Not very important, instead, seem to be the flushing discharge Q and the material composition (sand or gravel).

As the erosion gully in the reservoir progressively reduces its depth and steepness, sediment yield to the outlet becomes smaller and smaller, so that the average concentration decreases with the flushing duration. Experiments on the model have given the following expression (Figure 8):

$$\bar{C} = \frac{1}{\sqrt{D_*}} \frac{Y_0}{\sqrt{t_f}} \tag{5}$$

where the constant D* has the dimension of a diffusion coefficient with a value of about 0.5 m^2/s. The structure of Equation 4 can be justified by evaluating the sediment yield to the outlet through a simplified "diffusive approach."

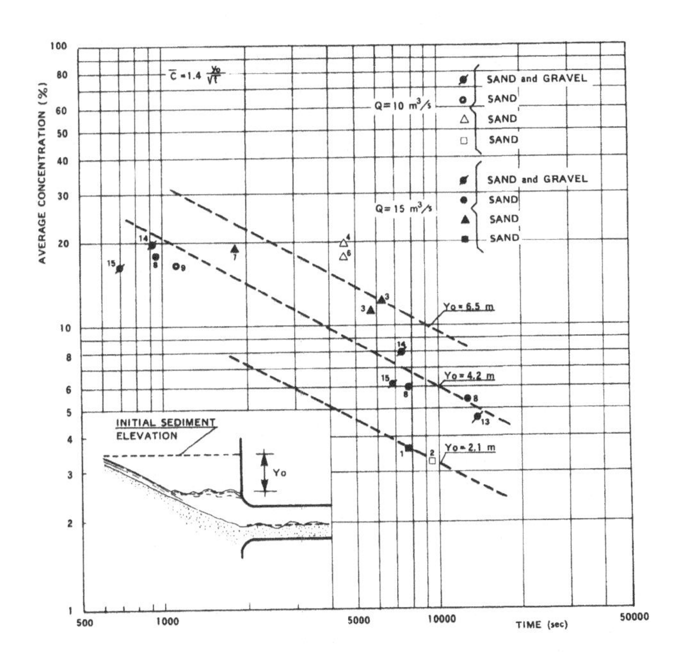

Figure 8. Transport of sediment in the reservoir. Time-averaged concentration depends on the duration of flushing and on the initial drop of water level.

On the basis of the model tests, the average concentration C has been also related to the hydraulic characteristics of the flow along the tunnel. As it was expected, the larger the sediment concentration, the larger the stream velocity should be, capable to convey the material. In controlling the sediment transport along the tunnel, however, the flushing discharge Q also seems to play a role. Experiments on the model have given the following expression (Figure 9):

$$\bar{C} = 0.1\frac{\overline{v^3}}{Q^2} = 0.1\frac{Q}{A^{-3}} \tag{6}$$

where $\bar{v} = Q/\bar{A}$ is the velocity of the flow averaged over the flushing period and $\bar{A}$ is the corresponding averaged wetted area.

As the flushing duration becomes longer, the value of $\bar{v}$ decreases (and the value $\bar{A}$ increases) following the variation of $\bar{C}$ with t_f (Equation 4). Note that the value of $\bar{C}$ is the "present" concentration $\tilde{C}$ averaged over the flushing period t_f; from Equation 4 one obtains:

$$\bar{C} = \frac{1}{t_f}\int_{t_f}\tilde{C}dt = 2\tilde{C}(t_f) \tag{14}$$

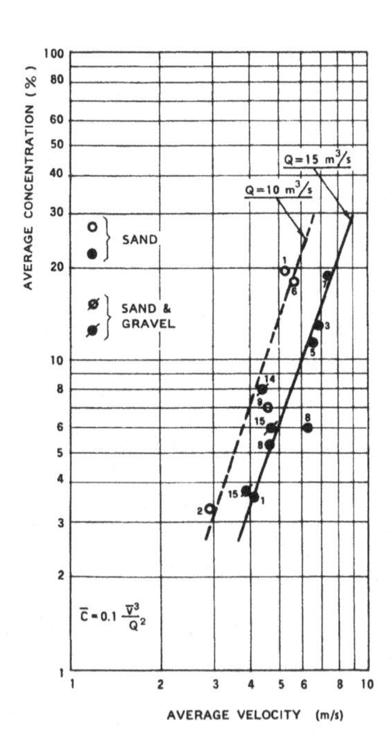

Figure 9. Transport of sediments in the tunnel. Time–averaged concentration depends on flow velocity and discharge.

Due to the intermittent character of the sediment transport, however, the present concentration $\tilde{C}(t_f)$ is not a real "instantaneous" value; in fact, the present value $\tilde{C}$ is the instantaneous value C(t) averaged over the time $t_d(<<t_f)$:

$$\tilde{C} = \frac{1}{t_d} \int_{t_d} C\, dt \tag{17}$$

corresponding to the passage of the sediment bulks (the so-called "dunes" mentioned in the previous paragraph). Deviations of the instantaneous values of C with respect to C, may explain the different structure of Equation 5, when compared to other expressions of sediment transports in pipes in stationary conditions.

6. CONCLUSIONS

The behaviour of the bottom-outlet during the flushing operations should be examined in two different situations: (1) quasi-steady regime, and (2) initial and final phases.

In quasi-steady regime, the tunnel is partially filled by sediments in such a way to adjust the cross-section to the flow velocity required to convey the incoming flux of sediments. In this condition both water flow and sediment transport are quite regular, except for occasional pulsations related to the presences of the so-called "dunes." Any possible intermittent clogging due to pulsations, however, is soon removed by a moderate temporary increase of the water level in the reservoir.

The initial and final phases are the most critical as far as the occurrence of permanent clogging is concerned. In order to avoid an inversible complete obstruction just at the start of the flushing operations, the outlet should have a sufficiently ample cross section (both of the tunnel and of the control gates), in relation to the tunnel length and to the initial thickness of the sediments above the outlet. As clogging may also take place as a consequence of the gate operations, attention should be paid to the proper sequence of opening and closing them.

Apart from the uncertainty arising from the possible effects of cohesion, a Frode physical model would be able to simulate all the above mentioned aspects, so that it can be used for the design of the outlet and for the definition of the flushing procedures.